EDUCATIONAL
SCHOOL GARDENING
AND HANDWORK

A handsome flower border   Chard Tatworth C. School.

EDUCATIONAL SCHOOL GARDENING AND HANDWORK

by

G. W. S. BREWER, F.R.H.S.

Inspector of Educational School Gardening,
Somerset County Education Committee

WITH AN INTRODUCTION

by

The Rt. Hon. HENRY HOBHOUSE

Chairman of the Somerset County Council

Cambridge:
at the University Press
1913

CAMBRIDGE
UNIVERSITY PRESS

University Printing House, Cambridge CB2 8BS, United Kingdom

Cambridge University Press is part of the University of Cambridge.

It furthers the University's mission by disseminating knowledge in the pursuit of education, learning and research at the highest international levels of excellence.

www.cambridge.org
Information on this title: www.cambridge.org/9781316509845

First published 1913
First paperback edition 2015

A catalogue record for this publication is available from the British Library

ISBN 978-1-316-50984-5 Paperback

PREFATORY NOTE

THIS work on Educational School Gardening is an endeavour to carry into practical school life the spirit of the following extract from *Jock of the Bushveld.* "Boys is like pups—you got ter help 'em some; but not too much, an' not too soon. They got ter larn themselves. I reckon ef a man's never made a mistake he's never had a good lesson. Ef you don't pay for a thing you don't know what it's worth; and mistakes is part o' the price o' knowledge—the other part is work! But mistakes is the part you don't like payin': thet's why you remember it. You save a boy from makin' mistakes, and ef he's got good stuff in him most like you spoil it. He don't know anything properly, 'cause he don't think; and he don't think, 'cause you saved him the trouble an' he never learned how! He don't know the meanin' o' consequences and risks, 'cause you kep' 'em off him!

An' bymbye he gets ter believe it's born in him ter go right, an' knows everything, an' can't go wrong; an' ef things don't pan out in the end he reckons it's jus' bad luck! No! Sirree! Ef he's got ter swim you let him know right there that the water's deep an' thar ain't no one to hol' him up, an' ef he don't wade in an' larn, it's goin' ter be his funeral!"

G. W. S. B.

1 *September* 1913

INTRODUCTION

I HAVE been asked to write a few words of introduction to this little volume, and I do so with pleasure.

We are all conscious now-a-days that there is a widespread dissatisfaction with our system of elementary education. It is said that by giving too much "bookish" instruction it is unfitting our children in the rural districts for their "natural future" on the land and encouraging them to seek for clerical employment in the towns to the detriment of our population as a whole.

This book may well be commended to those who sympathise with these views. It is perhaps through the medium of the school garden, the school workshop and the school kitchen, that we can do most to counteract the tendency complained of. Mr Brewer deserves the thanks of educationalists for the enlightened spirit in which he approaches his subject. He shows how, under wise guidance, the school garden may become the means of training boys in self-help, initiative, and the capacity for finding things out for themselves, and so learning the valuable lessons that only mistakes can teach. Mr Brewer believes that many school lessons, e.g., nature study, arithmetic, drawing, mensuration and composition, may be rendered more interesting and living by taking their subjects from the school garden. There is no doubt that progress in both literary and practical work is promoted by such intelligent co-ordination. Teachers should find the book suggestive and full of practical hints and ingenious devices.

There is, moreover, a still more important object which the author seems to have in view, namely, the formation of character in the training of boys in habits of industry, co-operation, and in what may be called the broad scientific outlook.

HENRY HOBHOUSE.

September 1913

CONTENTS

LIST OF ILLUSTRATIONS

1. THOUGHTS ON SCHOOL GARDENING

AIMS AND PRINCIPLES

THERE are many so-called school garden books in the market, but the large majority of these confine their attention almost entirely to the cultivation and raising of crops. The connection of the garden with the school is usually omitted or only briefly touched upon. It is, therefore, the aim of the present work to treat the subject from an educational point of view. Gardening operations, as such, will not be described although much gardening information will be given.

Personally I find that fewer teachers fail from lack of horticultural knowledge than from the want of knowing how and in what ways that knowledge may and should be used as a means of education. The real value of school gardening is, as yet, hardly appreciated, chiefly perhaps because it is looked upon, not as an educational handwork subject, but as a means of imparting knowledge of a country pursuit. There are thousands of schools without much hope, under present conditions, of having woodwork or other manual classes, to which school gardening could be made to appeal, provided the subject was put upon a different footing and had a different outlook.

From experience in teaching both woodwork and school gardening I can, without hesitation, maintain that

B.

school gardening may be made as valuable an educational subject as woodwork, and that, too, without much technical knowledge on the part of the teacher. As a matter of fact I find boys like gardening even better than woodwork. It is well, therefore, at the outset further to consider the aims and ideas that should underlie school gardening.

School gardening is not necessarily gardening, any more than cardboard work in school is box-making ; or woodwork carpentering. Gardening should be a school subject, taken as far as possible on the lines of other school subjects, notably woodwork. The latest principles of teaching woodwork applied to school gardening would help to lift this subject into its fit and proper place. It is in itself a most interesting subject and of utilitarian value. So far, the latter idea has been the chief aim and object of the teaching. In some quarters, however, the subject has been made unduly to correlate with the work of all classes and in all sorts of subjects simply for the sake of showing correlation. While it is important to realise how much of the school work can be interwoven with the garden work, the subject must not be taught for the sake of correlation. Such an aim would be extreme, and should be deprecated. Often, too, less is thought of the educational value of the subject to the children, than of what "the man in the street" will say, if perchance he look over the garden wall and see a line out of the straight, a boy holding a tool wrongly, a few weeds growing on a plot, or the least thing amiss.

Much may be learnt through our mistakes and hence a few things out of the straight, or a crop of weeds in a particular spot, may have, if rightly used, an immense educational value. Hence I would urge that in dealing with school garden work all idea of pandering to the man in the street should be ignored—rather, get him to take a walk round the garden with yourself as guide and then quietly point out some of the things that are being done. Schoolmasters, it has been stated, should be masters on their own quarter-decks. This applies particularly to gardening. A schoolmaster should not mind if he is not an expert at gardening; he should be an expert at teaching, and it is the teaching that is the thing.

It has already been mentioned that the methods of teaching woodwork might advantageously be applied to the teaching of school gardening, and, therefore, I will first go briefly over the ground dealing with this matter. In handicraft work, except perhaps gardening, the work throughout has to be the scholar's own production—he is told practically nothing, but by means of clever and skilful questioning he is led to discover things for himself, and further to describe and explain things himself, which previously he had little or no idea of doing. In other words the successful teacher of to-day adopts a suggestive method in dealing with his school subjects. He expects the pupil to take up the suggestions, develop ideas, discover facts, form judgments, and so make these things part and parcel of himself. What then the pupil has done for himself and of himself, he is likely to make part of himself for use on future occasions. It is, however, not the exact reproduction of an exercise worked that is of importance, but the means of finding out things for himself concerning an exercise. A child thus trained can be relied upon in other situations to exercise his faculties and work for himself

There are occasions when telling must be resorted to, when facts are dealt with which the pupil has no acquaintance with or no means of finding out for himself. Personally I find that there are very few facts that a scholar cannot be made to discover for himself. If this attitude of making the scholar rely upon himself is adopted with all the school subjects, it will not only develop self-reliance, but a number of other good qualities as well, among which we may name, observation, perseverance, judgment, a trained reason, and a knowledge of how to acquire further information.

To some, such a method of teaching as has been outlined may seem a little absurd, but I would say, give it a fair trial before condemning it. At first it may prove difficult for a teacher to avoid telling or showing the children every new step or alteration needed—and what is more, it may mean at first more labour and more care on the part of the teacher—but if persisted in, it will later be much easier and pleasanter for him and the results achieved infinitely more pleasing. For instance, suppose a drawing lesson is proceeding dealing with a dandelion leaf. Some teachers at once make sketches on the blackboard, and give full, very full, instructions to the scholars as to what THEY will see and how they should proceed. The result of such a lesson is more often than not a matter of disappointment, for after all the explanation and instructions have been given a large proportion of the class make fundamental mistakes in their drawings because THEY have not really seen—IT IS THE TEACHER WHO HAS SEEN. A method of teaching such as this is frequently employed lesson after lesson and in all sorts of subjects with the consequence that the children learn to rely upon the teacher to a very large extent.

A much better plan is to let each child have a leaf and then go round the class, talk to one scholar here, another there, asking each in turn pertinent questions— "Why do you do this, that or the other thing?" "What does this part represent?" "Compare this line or part with the original." "If this line on the object were continued, in which direction would it go?" "Does your line representing this go in a similar direction?" The scholar thus catechised is perforce obliged to look for himself and will thus often be led to see things which he never saw before. The teacher must not only make the scholar see things for himself but see also that he expresses them himself. The fact is that while the scholar is thus employed he is learning the HABIT of depending upon himself. It is said that " A bundle of habits makes the character." If the teacher can get the scholar to form good habits, he is helping in the great work of moulding the scholar's future welfare —a matter of supreme importance.

In taking a class by this method do not try to cover too much ground with each child at first, just give two or three suggestions and pass on to another child. This will enable the teacher to get round his class probably once or twice during the lesson. In a later part of the lesson, or on a future occasion, new steps may be added to the progress of each. What a scholar learns in this way he will be able to make use of himself in his next drawing lesson. In some schools marks are given for the finished drawing. These marks are usually awarded by the teacher. Where such is the case a change might be tried by making the scholars value their own work.

Thus suppose 10 marks to be the maximum for the drawing of the dandelion. A scholar is asked to say how many marks he considers his work worth, deducting, say, one mark for each pronounced fault or bit of bad work. The boy after some consideration thinks his effort is worth six marks. You then ask him for what he has taken off four marks and for this he must show his reason. If the teacher thinks the marks satisfactory, they are allowed. The value of this system lies in the fact that again the scholar has to depend upon himself, and further than that he learns to find out wherein lie his own weaknesses and faults, and these recognised there is hope of efforts being made to improve on future occasions. This method of marking should appeal especially to handicraft subjects, such as modelling, woodwork, etc.

A woodwork class may work somewhat as follows— after the first few lessons.—Before a scholar does any practical work he makes a drawing. This drawing should, in the case of ruler work, be to scale and English or Metric measurements may be employed. The drawing may be made in various ways so as to ensure a thorough understanding of drawings and their meanings. Thus, a model is given to a boy who measures it, draws a plan and elevations, etc. Isometric and oblique sketches or drawings may also be required of the model or of some part of it. Sometimes an oblique view of an object, dimensioned, is given to a scholar from which he is required to draw plan and elevation. At another time a sketch of plan with data of other measurements is given and an elevation asked for. The drawing completed, the amount and kind of material from which to make the object is written down. For some models a rough sketch of how the wood to be used may be set out to best advantage may be asked for. While the boy is making the model the teacher may question him concerning the wood he is using, the tools employed, precautions to take, etc. The questions naturally would be such that the scholar could answer from his own store of knowledge or observation. For example, suppose a piece of American White Wood is being used. The boy could tell the colour, freedom from knots, width of plank, and therefore from these facts, approximate diameter of tree, few branches low down, hence grown in a forest ; ease of working, nice finish, grain, annual rings. If the facts concerning the formation of annual rings are known, then by comparison with specimens of other woods such as yellow deal it can be seen that the annual rings are all about the same distance apart in width—unlike deal. This knowledge should show that American White Wood probably grew in a country where the amount of sunlight was about the same each year. A little questioning in geography would soon show that the wood is likely to be grown in America. Thus it will be seen that a very large part of the story of American White Wood has been developed from the boy's own observational and reasoning powers. A scholar so taught is likely to try to find out from other sources more about this wood. Work of this nature must make a scholar think—and think to some purpose too, and doing this gives him what may be termed "an attitude of mind." His judging as well as his observational faculties are well exercised. It is this attitude of mind that means so much in educating a scholar. The same principles can be applied with ease to school gardening, if only the work is undertaken with the intent to make it educational and not for the purpose of producing gardeners. School gardening may succeed to some extent even if worked on these latter narrow lines, but how much more might it succeed if conducted on a sound educational basis? The gardens exist for the scholars, not the scholars for the gardens. I shall endeavour in the following pages to put forth a number of suggestions, which I hope will appeal strongly to teachers, and so lead to school gardening (where facilities exist for its adoption) taking its fit and proper place as one of our leading educational subjects. There is no question of boys liking it, they simply revel in it. It is a " live " subject in which they can see tangible results and, therefore, they are the more keenly interested. It forms a break from the ordinary school curriculum and is on that account very welcome. Teacher and taught meet in a different atmosphere (often in more than one sense) and on slightly different terms. This change is beneficial to master and pupil alike. There is not the need for the strict silence of the schoolroom. Further, gardening is a sort of mutual work appealing strongly to the taught. While thus the boy is gaining much educationally in connection with his garden work he should at the same time gain much useful and practical knowledge of new methods and aims in culture and garden practice, so that he may go home and describe to his father how WE do it in the school garden, and the different results therein achieved, and hence by this means the gospel of the best ways of cultivation may be slowly but surely spread. The boy has an open mind, while that of his father is often narrow and fixed —for the father prefers to cultivate as his father before him cultivated—the boy by means of the school garden if properly worked will in course of time alter this state of affairs.

Another feature of school gardening is that it is likely to be an inducement to take up gardening as a hobby by many of the scholars in after school days, and if only a small proportion of the scholars follow up the subject in this spirit, it will have justified its existence as an integral part of the school curriculum. Too often, now-a-days, boys no sooner leave school than their whole minds are fastened upon sport—the watching and talking of the play of others. When this happens it is not long ere the boy deteriorates. The daily occupation of many scholars after leaving school is simply to attend to a machine, and thus in time they become narrow and mechanical: deterioration again. Nothing is better than to provide such youths with a healthy and profitable hobby such as gardening.

In the future development of small holdings school gardening may well take a useful place.

These then are some of the principles underlying this subject which it is well to bear in mind in connection with it.

The ideas and suggestions embodied in this book are not meant to be rigidly followed—they are intended to give lines of thought so that school gardening becomes a schoolmaster's subject rather than a gardener's subject.

" It is the spirit that giveth light, the letter which killeth."

2. "A BEGINNING"

THE school garden should be situated as near the school as possible. It will thus be much more useful and valuable than if at a distance.

The size of the garden cannot always be regulated, but if possible it should contain sufficient land for plots for the scholars : also a Common Plot, an Experimental Plot, a Fruit Plot and a Flower Border. The main consideration of course will be the plots for the boys. These should be, as far as can be arranged, long and narrow, say 10 yards by 3 yards. This shape allows a larger number of rows than is possible with a piece of ground of shorter length and greater width. One or two boys may work each plot. If two boys work a plot, then a senior and a junior boy might well work together. Fourteen boys working on seven plots can be supervised better and with greater ease than the same number of boys working on separate plots.

A plan of a school garden as laid out under the direction of the writer is shown. This garden was measured, pegged out, paths made and edging put to the main pathways by the boys themselves. The work of laying out the ground would not often come in the life of the school, but when it does come I think the scholars should take their share in it—it thus becomes essentially "their garden."

The Common Plot, as its name implies, is a piece of ground set apart for work in common or for practice work. It may be used for both purposes. For instance, celery and runner beans may not be grown by

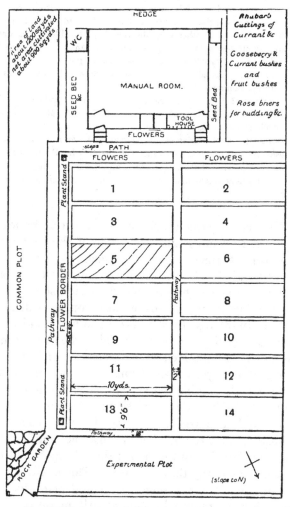

Plan of Nailsworth Boys' School Gardens, 1910

each individual scholar but a row or two should be cultivated on the Common Plot. It is not advisable to plant too much winter green stuff on the boys' plots or the ground will not be available for winter trenching, etc. Hence this is another crop for the Common Plot. Practice in planting small seeds may be given on this land. Plants specially required for the Nature Study lesson may also be grown here. Some of the less common vegetables, such as Celeriac, Salsify, Endive, should be grown on the Common Plot.

Experimental Plot. If the garden is large enough it is a good thing to set apart a portion of ground to be utilised for simple experimental work, such as will be described later.

Fruit Plot. Each garden should have at least a corner where cuttings of gooseberry, red and black currants may be raised. Room might also be found for growing rhubarb. Grafting and budding of fruit trees and budding of briers might also be included by those who can undertake such work.

The school garden year of the Board of Education begins in November and this season of the year affords an excellent time to make a start. The first work naturally will deal with the garden as a whole, and a drawing to scale of the garden and its parts should be the first consideration. To make this properly will necessitate a knowledge by the class as to how and by what means the garden can be measured. The desired information can be obtained by a discussion with the class. Tape measure, yard stick, Gunter's chain, etc., may be suggested as suitable measuring tools. If a Gunter's chain can be had, then this should be brought to the school and a conversational lesson given concerning it—dealing with the number of links, the length of each link, and therefore of the chain; the tabs and what they represent and why they are used. The flexibility and strength of the chain should be noted and why these properties are required. Mental arithmetic questions should show the connection of a chain with yards, rods, furlongs, acres, etc. Do not tell the class facts of this kind, rather make them measure and find them out for themselves. In the drawing lesson sketches may be made of individual links, of the tabs, and of the end link. The handle link may also be sketched. In the composition lesson a description of the chain and its uses may be written. The mensuration dealing with finding the areas of right-angled triangles and rectangular figures may be revised if necessary. It may be pointed out that when land is measured it is generally, and as far as possible, mapped out into plots that are either rectangles or right-angled triangles. Thus prepared, the class, including scholars who do not take gardening, can go to the school garden. Each scholar should be provided with a sheet of drawing paper and a pencil. A description of the garden and its chief parts should be given by the scholars, after which each should make a rough sketch plan of the garden, on which will be written down the dimensions as determined. Various members of the class will do the measuring, and as each result is obtained it should be written down on the sketches. When the garden has been thus measured, the various parts will next be dealt with and these it may be found more convenient to measure with a tape measure or yard stick. All measurements as found are written on the sketches. The work in measuring thus far indicated may well occupy more than one lesson. When the class has made complete dimensioned sketch plans the work will be transferred to the school where drawing to scale from the sketches should be made. Discussion with the class will first take place and the scholars should suggest what scale should be employed. Here I would say that when a child suggests a certain scale—whether a good one to use or not—go into the matter with the class and let the scholars themselves decide if it is suitable and why. Work done in this way gives value to the subject. Never mind the time it takes. The work is not so much to make a scale drawing as to educate the child, and if this is the aim then time is not the important factor it is often thought to be. Having at length settled upon a suitable scale, the position of the drawing upon the paper may next be briefly considered. The teacher should not draw the plan on the blackboard line by line, telling the children what to put down and where. Rather question them as to the length of the line and its relationship with the drawing. Then draw the line freehand on the board and let it be very carefully drawn by the scholars. By sketching on the board instead of measuring and ruling the teacher will have an opportunity of watching the work of the scholars and dealing with any difficulties that may arise with individual pupils. Needless to say no dimensions should appear on the finished drawing. Each garden boy might shade or colour the part that represents his plot and show it in relation to the garden as a whole. As a general rule a scale need not be constructed on each drawing before the drawing is made. If the work has been properly understood, "an inch to a yard" or "$\frac{1}{36}$ size" is sufficient, as the rulers provide all else that is necessary. It is usually a waste of time to construct a special scale for each scale drawing. Dimensions ought not to be put on the finished drawing, for the following reasons. A drawing to scale is supposed to be made in accordance with a scale stated on the drawing sheet, which gives the key to the lengths of all the lines. If dimensions are written down, there is no need to adhere to the scale. Doing away with the scale is doing away with accuracy—one of the very things a scale drawing is intended to develope. In scale drawings with dimensions written on, small inaccuracies would make no difference, while if undimensioned, mistakes would appear in the finished article. One other strong reason for not figuring the drawing with sizes is that a boy should be taught to translate a scale drawing back into full sizes. A person learning shorthand not only learns to write the characters but also practises changing the characters back into words again. So it should be with scale drawings, and exercises in this should be given at times. In after school days the scholars will have to work from scale drawings much more often than they will have to construct such drawings.

The next work may well deal with mensuration problems, such as finding the area of the whole garden. It is as well to express the area not only in standard terms, acres, square yards, etc., but also in terms of measurement peculiar to the neighbourhood, e.g. lugs, etc. The area of the paths and the separate parts of the garden may be found and also the percentage of land occupied by the paths, by a boy's plot, the flower bed, etc.

Another lesson may be devoted to measuring the garden again in the metric system. It may be necessary to make a metric rod for this work. A scale drawing may then be made using metric measurements. This is practical work that will repay doing. Areas, etc. may also be found as before and comparisons made between the English and Metric measurements. Printing exercises might be given in fancy or other lettering for the headings of the scale drawings. The scholars should draw to scale the common plot, the experimental plot, the tool house and an inside wall, showing positions of tools when stored, also the tools and appliances. The drawing to scale of the individual plots will be dealt with later. Thus it will be seen that the garden may provide sufficient scale drawing for a year's work in this subject. The measuring and the drawings would be made either in the garden lesson time or in the arithmetic lesson, preferably the latter.

3. "MY" GARDEN.

"In the 'I made this' or 'I made that' is the making of 'I.'"
H. HOLMAN.

GARDEN work, if properly taught, should lead a scholar to take a pride in himself and his handicraft, and therefore the more a scholar is allowed to do on his own initiative the more of the making of "I" in the process. How often does one visit school gardens and find that every individual plot is cropped exactly alike in all respects! If such gardens were visited when planting operations were in progress, it would be found that the boys were told exactly what to do and how to do it. Two boys would for instance manage the placing of the line across as many of the plots as possible and then each boy would do exactly like his neighbour. It can readily be seen that if the teacher is taken away for a while the gardening would cease, as the work is done "to order." Surely it is only a "telling lesson" over again !   Where the school garden work is conducted on these lines great opportunities are lost.   The spirit of the manual room—of independent and individual effort by the scholar—needs to be brought into active usage in conducting the school garden operations.   This may readily be done.   Let each boy plan how he would like to plant his plot.   Before he could do this he must have some acquaintance with the crops he wishes to grow. Provide the class with a number of weekly gardening papers, old ones will do, and instruct each boy to select a firm of seedsmen to whom he would like to send for a seed list with a view to buying seeds.   The boy should settle for himself whether he would use a postcard or a letter for this purpose.   Those who elected to send a postcard should draw in their books or on their paper, unruled for preference, two rectangles the exact size of a postcard to represent the front and back sides.   The one side should have the address of the firm written on, and in its proper place a space should be ruled to represent a stamp and inside this the value of the stamp should be written.   The other side of the postcard should have written on it a request for a seed list, together with name and address of the sender.   Those who preferred to send by letter would first write in proper form their request and then rule a rectangle to represent an envelope, which would then have address and stamp put in their right places.   It is surprising what useful knowledge work of this kind will bring out and also what ignorance exists of such simple matters.   Two or three requests may be written and actually sent through the post.   The copies of seed lists on arrival together with any others obtainable may be handed to the scholars with instructions to look them through and then make out a list of seeds, etc. for their own gardens. The amount and cost of the seed should be put down, and finally the grand total cost of seeds for the class should be made. Here, then, will be a most favourable opportunity for impressing the need and value of co-operation in the purchase of seeds—a lesson that cannot be learnt too early or too thoroughly. Head teachers might with advantage be allowed a sum of money for each scholar for providing seeds, or alternatively the Education Authority might either issue a seed list of their own, or else allow choice to be made from the seed lists of certain firms. At the present time, under some authorities, a similar supply of seeds is sent to each school, regardless of the requirements : too much of some kinds, too little of others, unsuitable varieties for the district and kinds not required because of seed saved from last year. If a master could purchase his seeds the question of cost of seed and co-operation would not be simply "playing at ordering" but would be the real thing, and it is wonderful what an effect the real thing has upon child life. Even if the cost of seeds thus supplied were a little greater to the Authority, the result would justify the expenditure. One other point, too, might be noticed about this method of supplying seeds, and that is the need to keep accounts. In fact where there is any buying and selling, a simple income and expenditure account should be kept. The next step, whether "making believe" or the real thing, would be to make out one seed list and write for the seeds, enclosing the value for same and giving directions how and by what routes the seeds are to be sent. Each child should draw up a suitable letter and state the sum of money enclosed and in what form it is sent. This would provide scope for dealing with the different forms by which money can be sent, and precautions to be taken regarding the safety of the remittance. Only deal with mistakes after they have arisen. Too much importance cannot be attached to studies of this nature which would be taken in the composition lesson.

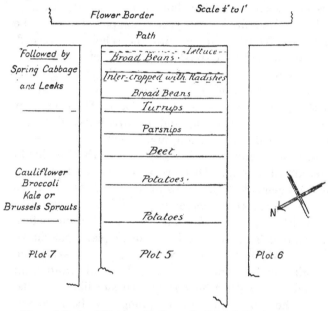

Plan of Part of Plot 5. Nailsworth School Gardens, 1910

When this stage has been reached attention should be directed to the planning of the boys' plots. These should be measured both by English and Metric systems, and scale drawings made showing the position of the plot in relation to the neighbouring plots.

A draft plan should next be prepared by each boy (if two boys work on one plot they would do this jointly)

showing exactly how HE would like to plant his ground.
On this copy should be written the names of the crops
to be grown, the distances apart of the rows, and the
position of each crop.   Each boy would plan his cropping
as he thought fit.   It is thus possible for 14 boys to have
14 different ways of cropping their ground.   At first
this may seem confusing, but such is not the case in
reality.   Each draft plan when completed should be
gone over by the teacher with the scholar concerned,
and any alterations arrived at by means of questioning.
For instance, a plan shows a tall growing crop
planted in front of a low growing one.   Questions
on the position of the sun in respect of the crop at
early morning, at noon, and at evening and therefore the resultant position and amount of shade at
those times would lead the boy to see why he should
make an alteration.   Distance apart of the rows, and
the arrangement of the crops in some kind of order are
the most necessary corrections.   Whatever the boy
has written down, make him give his reason before
condemning him.   I once had a draft plan brought to
me from which carrots were left out.   On asking the
boy why he had omitted carrots from his draft plan
I found that neither he nor his parents liked carrots.
When the draft plan of cropping has been finally
approved it is drawn to scale on the scale drawings
of the plots previously prepared.   Each boy will plant
his land entirely from his own scale plan and according
to the cropping shown thereon.   He not only makes a
scale drawing but he learns to work from it.   Of course
no measurements will be written on this plan.   If follow-
on crops are shown on the plan as in diagram above,
then the complete cropping for a year may be shown.

To obtain practice in the sowing of seeds, etc. some preliminary work is undertaken on the Common Plot. The boys themselves do this work, taking shares as far as possible in drawing the drills, sowing the seeds, etc. They are asked which tool to use in making the drill, and if more than one is suggested let each be tried and the question decided upon its merits. They are questioned at the time what distance apart the rows should be and why—as a general rule crops are planted as far apart as the plants grow tall. The depth of the drills and special points concerning the preparation of the seed beds should also receive attention, as they arise. For instance when sowing onion seed the reason for making the ground firm should be discussed. (An onion produces a bulb which should be mostly on the top of the soil, while loose ground tends to produce "necky onions.") Great care should be taken to have the distances apart of the rows measured correctly, and after placing the line in position the distance from the last row should be tested by measuring in one or two places. By this means the boys develop accuracy and learn to recognise that they must do well all they have to do. "Anything-will-do" sort of work must never be allowed. From time to time the boys are given a few seeds to examine and describe as to shape, size, colour, etc. Thus they are kept keen all the time ; there is always something fresh for them to discover. The using of the tools in drawing drills, the handling of the seeds, and the covering in of the drills all need careful watching. Let the boys point out any faults they notice as the work is proceeding.

Thus prepared the boys could be trusted to do

their own plots, but before this work is commenced the seeds should be divided amongst them. The most satisfactory way to do this is to prepare beforehand a number of seed packets in various sizes, sufficient for each boy to have one for each kind of seed he will have. If paper modelling is part of the school curriculum the making of these will be a simple matter, otherwise the class may dissect a seed packet and determine how to set out a similar shape on paper and how to put it together by creasing, gumming, etc. Alternate sizes and shapes might be suggested. If cardboard modelling is done in school, then boxes large enough to hold each boy's seed packets could also be made. In the alternative the boys would possibly provide tin or cardboard boxes for themselves. If a balance is available the seeds might be divided by weight. The name of the seed should be written on each packet. These preparations having been completed and the plots prepared, the planting may be begun. Each scholar takes his scale drawing with him into the garden and also a rule and garden lines in addition to his seeds and the requisite tools. He can then commence to work his plot in accordance with his planning. He measures the distance, sets his line in position, and draws a drill or makes a trench, as the case may be. Here I would mention how important it is as a rule for the boys to get into the habit of walking round their plots by the paths to get from side to side, rather than by walking over the garden. When a boy has his drill ready it should be inspected by the teacher before the seed is sown. If the lines are not put in at the correct distance on each side, the boy should measure them again while under observation and remake the drill. The boy may be asked how he intends to sow the particular

seed, what distance he will have to measure for his
next row and where and how he gets these particulars.
Other matters relating to the work may also be given.
The teacher will thus pass from boy to boy, observing
how the work is done, correcting any faults in the handling of the tools or questioning on any points that may
call for notice. The boys will get their measurements
and work entirely from their plans. It is a good idea to
have the rows labelled with name of crop and date
of sowing, or else to mark these details on the plan.
Throughout the year the boys will do independent work
on their plots. Let them suggest what they propose to
do and question them for their reasons. All through,
it is the boy, not the garden, that is the main consideration. I venture to say, however, that the garden does
not suffer in the least for being worked on individual
lines. The working from plans is only doing in the garden
what is done in the manual room when the scholars
make objects from individual scale drawings. This
method of work creates independence. It allows, as
the crops spring up and mature, of interesting contrasts
and comparisons being made. Above all else it forms
character and that of a right type. In after school days
the youth will know how to depend upon himself. It is
the making of " I."

4. DESCRIPTIVE WORK

THE garden can and should provide material for
other school lessons than those already dealt with,
notably drawing, composition, and nature study. Drawing is supposed to be entirely done direct from natural

and other objects and hence the school garden can
be of the greatest assistance in providing varied and
suitable material, but it is a matter of surprise how
often the garden is overlooked for this purpose. Some
of the common weeds found there make excellent copies,
and can be obtained nearly all the year round: for
instance, shepherd's purse, groundsel, dandelion, plantain.
If a complete plant, including roots, is too difficult, then
the plant can be drawn in parts, a leaf, blossom, seed
pod, roots. The drawing of the roots should by no
means be overlooked. Blossoms and pods of runner
beans, broad beans and peas make good examples,
while a "*chitted*" potato, an onion, leek, carrot, etc.
may furnish other copies. Seeds, e.g., broad beans,
kidney beans, pea, parsnip, etc. will be useful and
may be drawn both actual size and enlarged.

A series of drawings of parts of a plant at different
stages of growth would be attractive and provide useful
work. For instance a broad bean seed could first be
drawn, and then again when it shows the first stage of
growth. Another drawing would be made when the
seed leaves appear. Further sketches would include
the bean when the leaves are a little more developed;
a full-sized leaf; a blossom—a longitudinal section of
a blossom might be made also. It should be noted in
what manner the flowers are fertilised and drawings
made of the creature or creatures which do this work.
If the flowers are examined it is possible a number of
them will be found to have a small hole at the base
of the corolla, and efforts might be made to find out
how and for what purpose the holes were made. A
sketch of a growing bean plant could be made by the
more advanced scholars. A bean pod; a bean pod

opened ; a ripened bean pod would complete the series of drawings. The value of such work lies in the fact that the plant is under continued observation for several months of the year. For object drawing, the garden tools, such as the fork, spade, trowel, dibber, garden basket, and others may be used. The Dutch hoe, although a simple-looking object, often proves a stumbling-block and very funny attempts are sometimes made in trying to show that the blade is set at an angle with the handle. Where this is the case a second view taken from the side should overcome the difficulty, and in this case if the one view is properly projected from the other another form of drawing would be taught.

If the class can do advanced work in drawing they might go into the garden and make studies of growing plants, or make rough sketches on the spot and afterwards draw the plant from memory in school. The small animal life of the garden would make a very interesting change from plant studies. Woodlice, earwigs, a butterfly grub, chrysalis of a cabbage butterfly, a cabbage white butterfly, a centipede, a "daddy-long-legs," garden spider, an ant, snail, slug, ladybird and others may be drawn. If these are drawn larger than natural size, as is advisable in the case of the smallest animals, then alongside the drawing should be drawn to scale a cross—simply two straight lines, crossing each other at right angles. The upright line being the exact length of the creature, while the horizontal line gives the width across the body, or across the expanded wings. A set of diagrammatic drawings could be made representing such gardening operations as trenching, double digging section of a drill, or celery trench ; method of placing kidney bean or pea sticks. The flower bed can be made to furnish an endless number of copies. Drawings of currant and gooseberry cuttings and of fruit may be obtained from the Fruit Plot. This work would of course be done in the ordinary drawing lessons. There is no need for each member of the class to be working from the same kind of copy. In fact it is better for the work to be done on as individual lines as possible. Thus it will be seen that the garden can furnish a large and very varied assortment of copies, suitable for scale drawing, sketching, the making of diagrams, memory drawing, and the drawing from objects and natural forms—truly a wide field.

Composition is one of the most important of school subjects, and anything that will tend to make these exercises of a better and higher standard must be called into service. Nothing will do this better than plenty of essays on gardening subjects. The children thus write about things with which they are familiar and about subjects in which they are often most keenly interested.

The writing of postcards and letters has already been referred to and there are often other times when similar work can be done. For instance in early spring a supply of lettuce plants may be desired. From a gardening paper a suitable advertisement offering lettuce plants for sale might be selected and written on the blackboard and the class instructed to apply for a certain number of plants, to deal with the enclosure of remittance and to state how and by what route the plants are to be sent. Again, someone interested in the school may give stones for a rockery, manure for the garden or plants for the flower bed, etc. Letters of thanks should be written as a class exercise and one of the best letters selected to be copied and sent to the donor. Examples such as these deal with actualities and help to bring life into the school. For essay work practically every subject mentioned as being suitable for the drawing lesson may be requisitioned as a topic about which to write. In many cases the actual object will be beside the scholar as he writes and he will have to describe it. Descriptive work of this nature, if properly superintended, must lead to accuracy of observation and expression. It should be insisted upon that exactness will be expected.

Thus I suggest a close and intimate connection between the drawing and composition lessons, for both deal with expression work, founded upon observation and knowledge. Do not let all the children write on the self-same subject; give each a separate subject. The scholars are thus bound to do individual work. Do not let the class copy from rough notes, or from elaborated notes or other "pegs" of like nature. The independently worked composition may not look so nice at times as that where a certain amount of assistance has been afforded, but it cannot be borne in mind too much that the composition is given to educate the child and not to produce model work, or a model book. Enough regard is rarely paid to this fact. One other point, the exercises should be promptly and most carefully marked and the mistakes corrected. Further subjects are methods of planting, and of raising certain crops and other matters relating to the cultivation of the garden. If a child is set to weed the garden path he might in the composition lesson be asked to describe which weed he found the most troublesome and why, and also to state any steps he would take to try to get rid of it. A garden bonfire might form another suitable subject to write upon. Here not only should the composition be descriptive of the fire but the value of the residue in form of ashes should be noticed too. It is seldom that the garden tools are talked about—their shape, size, fitness for the work they have to do, materials of which they are made, where the tools are made, their approximate cost, the early forms of some of the tools, mechanical principles involved, etc.—and yet this information should be as useful to a boy who takes part in gardening as it is for a boy in the woodwork class to be able to describe a chisel or a plane. Descriptions of some of the seeds, such as parsnip, onion, etc., might be asked for at times, while occasionally a boy might be given a few different kinds of seeds, unnamed, and be asked to name each lot and state his reasons for his selection. Never mind if mistakes occur in doing this exercise, for the point is that the children have had to concentrate their attention on some seeds and to think about them, and in future they will be careful to note of any seed its shape, size, colour, etc., all of which train in habits of observing. One complaint frequently raised is the amount of time this work takes from that allotted on the time table. No extra time is demanded seeing that the exercises worked are in lieu of others, and while the subjects are on gardening topics yet it is composition and not gardening that is being done. Naturally other types of composition exercises would also receive attention. The garden essay might be taken once a week, or inter- mittently as occasion arises—but it should be taken.

5. NATURE STUDY

NATURE study should be intimately connected with the garden and the scheme of lessons therein, and should be such that the lower part of the school would deal with the simple aspects of those things which they would work out in more detail in the upper standards. Wherever possible the lessons should be taken from actual objects, which should be so used that the children would be led to observe correctly, to draw what they see, and describe the same in words. It is not my intention to give set schemes, but rather to indicate what subjects may be called upon for lessons and the manner in which the lessons should be handled. For the lowest group choice should rest upon simple studies of plant life, of small animal life, simple experimental work in plant growth, soils, and weathering agents, evaporation, condensation, clouds, rain, heat, cold, etc. The second group could carry on the work of the first year to a more advanced stage, and in the third year more detail could be added. The method of conducting the nature lessons is of the highest importance. Suppose then our lowest class has a section devoted to plant study. What should be done? In the first instance the class should examine a number of plants and say of them that each consists of root, stem, leaves, flowers, etc. The lesson might take the form of a chat, the children being encouraged to talk about the plants; the teacher by skilful questioning might keep them to the points with which it is considered wise to deal. In the drawing lesson sketches of some of the simple parts of a plant might be made, while in the composition lesson simple sentences descriptive of the plants should be written. It is essential to get children from the first to express their experiences.

A further lesson might treat of the roots, another the leaves and their shapes. The leaves might be sorted out by the children into forms that are somewhat alike in shape, and thus they would have to exercise their judging faculties. A collection of leaves might be made by the children, the specimens being mounted and kept throughout the year. The blossoms of flowers could also be studied. The names of the parts of the blossoms can be told as the lesson proceeds, if it is deemed advisable to do so, but too often such lessons develop into names and nothing else. It is in seeking to drive home "names" that teachers are apt to get far away from the true spirit of nature study. The names are not the essential features. It is the attitude given to a child of looking correctly at, and understanding things that is important. Knowledge, obtained by observing, and ability to express should be the aim. Growing of peas and beans, etc. from seed could be carried on in the school and would form interesting lessons. Following up the study of plant life in the next higher group the work might deal with a general outline of a few plants, such as the crocus, onion, dandelion. The class might make a collection of seed pods, to show how seeds disperse themselves. Another collection could be made showing how plants defend themselves. The habit of collecting is quite a good one if carefully watched and regulated. A child should be accustomed to write down where he found a certain specimen, whether in the open field, in the lane, or the hedgerow. Some more simple work in plant growth should be conducted as part of the work, e.g. growing of bulbs, wheat, etc. If plant life is made a speciality in the school, then the plants of a hedgerow, or of a pond and pondside, could be studied. There is no need, nor is it desirable, to go into too much botany in connection with the subject. If this kind of work has been properly handled, the upper group could undertake the study in rather more detail of the weeds of the garden or of a meadow, etc. Side by side with these lessons a course of simple experimental plant physiology would be valuable as tending to scientific training.

Perhaps no part of nature study is more attractive to the scholars than that dealing with small animal life. Children revel in things that have " go " or motion. In the lowest group in school such things as a butterfly, a spider, a mole, etc., may be dealt with. If possible take the class into the garden and search the cabbage leaves until one is found on whose underside are eggs of a butterfly. Let the children look at the eggs through a magnifying glass, and describe what they see. Let them suggest why the eggs do not fall off. Let them rub some off and then without any adhesive try to stick them back in place. Find another leaf or two on which there are eggs and mark these for future observations. Each school session a child might be sent out to look at the eggs to see if any change had taken place—this excites curiosity, an important adjunct in education. In time the grubs are hatched. Again an examination should take place as before. The children should state what the little grubs are doing and it should be pointed out how wonderful it is for these tiny little things to be able to look after themselves. They may then be

watched day by day living on the cabbage leaves, and
any changes noticed. If a child ventures an opinion,
do not say "That is wrong" or something to that
effect. Rather appeal to him to describe why he
ventures the opinion and then ask other members of
the class whether they agree or not, and if the question
cannot thus be decided the teacher might give his ideas.
It is the child all the time that must be kept in mind.
He must be made to see, to reason, to judge for himself,
and this he cannot do if he is continually told things.
The grubs as soon as large enough can be described as
to number of legs and it does not greatly matter at this
stage if the pro-legs are called legs ; but a child can
easily be led to see that one set of legs is for walking
and the others for clasping. Let them note which part
of the leaves is eaten and state why. In time the
grubs will be fully fed and will be leaving the plant to
turn into chrysalids, and if there are railings or a fence
near some of the grubs will most certainly be seen
climbing up these and chrysalids will also be found
thereon. It will probably take too long to wait for the
chrysalids to turn into perfect insects to continue the
investigation in the garden. The motionless chrysalids
can be examined and wonder excited in the child as to
how they changed thus and for what purpose. Old
skins can generally be seen and they and the manner
in which the chrysalid is suspended can be described.
That it is still alive may be determined by touching it
when it will wriggle about a little. It may be necessary
here to tell something of the chrysalid stage as one
wherein the creature keeps quite still and stationary—
no food, etc., but a most wonderful change is taking
place nevertheless, and that in due course will come a

butterfly. The choice of railings, etc. for this period of its existence can be commented upon and reasons for it advanced. The study of a butterfly thus far will probably occupy a few weeks, but the visits to the garden need not occupy more than a few minutes in some cases.

Some other subject in nature study could probably be undertaken to fit in with the somewhat disjointed study of the butterfly grub. In the garden it should not be a difficult matter to watch cabbage butterflies flitting about, and children could note among other things that they do not themselves eat the leaves, but rather that they visit the flowers. If a cabbage leaf is visited by the butterfly, it should be noticed where it settles and after it has gone the leaf might be inspected with a view to finding thereon a cluster of eggs. To complete the life-story of the butterfly it may be necessary to catch one and take it into school and there study it. The main part of the work however has dealt with living things and thus the curiosity has been aroused to the utmost. A lesson such as thus indicated is valuable because something more than facts concerning the life-story of a butterfly have been observed, the foundation has been laid of a habit of looking with a purpose. This then is of much more importance than the facts gathered.

Needless to say drawing and composition should be intimately connected with these lessons. The work of the intermediate classes can be carried on along similar lines, and for a change the lives of pond creatures might be studied by visits to ponds and by keeping some of the creatures in a school aquarium. There are of course plenty of subjects that could be selected from the garden, as for instance, a garden spider, an ant, a wasp, a slug or a snail, etc. But it is well to bring variety into the work, as the upper classes might devote their attention mainly to the creatures found in the garden. The study would be made more individualistic than in the lower groups. Few formal lessons would be given but the scholars would be encouraged to take a keen and active interest in the different forms of life met with in their garden work. The kind of procedure that is usually adopted is somewhat as follows:—In the course of digging a boy finds a grub, or a beetle. Often his sole desire concerning it is to find out its name, and if told this further interest in it is at an end. If questioned as to what he intends to do with the specimen he will probably reply that he proposes to kill it, heedless whether it is friend or foe. Instead of dealing thus with specimens found it would be very much better if the scholar were told to write a description of it, and make a drawing or drawings of it, and when these had been completed satisfactorily, to find out more of its life's story. If the grub is found in the soil the scholar should endeavour to keep it under observation in conditions as similar as possible to those of nature, and for this purpose he should be led to suggest keeping it in soil, say, in a glass jam-jar, or box. Note should be taken of what food the grub was likely to have been feeding on and some of this supplied to it. A little damp moss put over the soil will help to keep it from getting too dry. During the time it is under observation note should be kept of its doings and when a change occurs by its turning into the pupal stage, another written description and drawing should be made of it. The chrysalis should be put back into the soil, which should continue to be kept slightly damp by means of the moss, until the perfect state of the creature's life has been attained. When that happens a further drawing and written description should be made, and if the name of the creature is not known then hand the boy a book or books dealing with similar forms of life and let him look through these and discover, if he can, the identity of the creature and more about it. Let him give his reasons for his decision. For the purpose of reference the Leaflets, in book form, issued by the Board of Agriculture are admirable. In this method of dealing with the creature it will be noticed that the boy does all the study for himself. Teach him this lesson of self-reliance thoroughly in school and he will have learnt something that will be serviceable to him through life.

Another method of study can be developed as follows. Perhaps it is noticed that some of the growing onion plants are turning yellow in the leaves and have a sickly appearance. Careful examination of such specimens will very likely reveal the presence of maggots attacking the bulbs. If so dig up an onion that is affected and also dig up the soil for two or three inches from below the onion. Probably in this soil from underneath some brownish looking chrysalids may be found. Place some of the soil and the chrysalids in the bottom of a small glass jam-jar and then put in the onion plant with the rest of the maggots. Cover the top of the bottle securely with paper pierced with a few tiny holes. Write date on the cover when specimens were collected. Keep daily watch on the jar and note the date when the chrysalids become flies. The time for this to happen, if in early summer, is usually about a fortnight to three weeks. Hence several interesting channels of thought

3—2

are opened out, e.g. What would have happened supposing the specimens had not been disturbed? How
many broods are possible in a season and how do the
latest broods survive the winter? What is the approximate number of descendants that might be produced in
a season from one pair of flies? What natural enemies,
if any, have the flies? Also enquiry should be made as
to how to circumvent the flies by finding out as much as
possible about the onions affected, the times of sowing,
the varieties chiefly affected, kind of soil, loss due to the
flies, the necessity of making sure that all maggots and
chrysalids are destroyed under each onion affected, how
a neglected garden may infect good and carefully kept
gardens near. Thoughts can also be directed to the
value of onions imported from abroad and how, if we
could improve conditions of growing, this amount could
be greatly decreased. The grubs, the chrysalids, the flies,
the affected onion, may all be drawn and described. To
make the drawings it is necessary for the scholars to
use a magnifying glass. There are many examples of
a similar nature that can be studied in the same manner,
as for instance, gooseberry saw-fly, raspberry shoot-moth,
currant clearwing, etc. A study of beneficial creatures
in the garden is most interesting and likely to prove
useful too. In this connection the life-stories of ladybird, gauzy-green lacewing fly, hover fly, ichneumon
flies, worms, centipedes, etc., may well be investigated.
The birds that visit the garden may be watched, especially with the view to finding out the nature of their
food. The value of seed-eating birds in feeding on the
seeds of weeds should not be overlooked. A collection
of snails found in the neighbourhood might be made and
thus the boys' fondness for collecting fostered. Habits

such as these often grow into hobbies. Specimens obtained should have notes written about them, stating nature of surroundings, such as hedgerow, dry wall, tree, etc. ; what feeding upon, etc. It is a good plan for the garden boys to take with them into the garden a tin can or a bottle into which they can put their captured grubs, etc. A cigar box could easily be made into a useful collecting case by having a sheet of thin cork or cork lino glued to the bottom. A killing bottle might, too, at times be found serviceable. The preservation of specimens is fully dealt with in the chapter on the School Museum (below, p. 56).

Work in connection with soils, their nature and formation, effect of slope, shelter, etc., weathering agents, etc., may also be undertaken. Simple soil maps of the district may be made. In connection with soil work it might be useful for the scholars to be able to make a rough analysis of their garden soil. This is not a difficult operation and can be carried out in the following manner. Obtain a spit of soil that is characteristic of the garden. The spit should be dug out about nine or ten inches in depth. Place this soil on the floor and thoroughly mix it, after which allow it to dry. A small sample is carefully weighed and then burnt in a suitable vessel—an iron spoon will answer the purpose very well. Note the strong smell at first given off by the burning soil. After it has been thoroughly burnt and no further smell is given off, the soil should be again carefully weighed. The loss of weight is due to the removal of organic matter or humus in the soil. Another portion has next to be sorted out into different degrees of fineness and the proportion of these grades determines whether the soil is heavy, light, porous, etc. This part of the

operation is done by washing the soil with water. Place
the soil in a large jam-jar or other convenient vessel,
add water and then stir thoroughly. Pour off the
muddy water into another glass vessel and allow it to
settle. The deposit obtained contains part of the clay
and silt in the soil. Continue to wash the soil until the
water runs off quite clear, leaving a gritty deposit.
Dry this latter and weigh. This represents the sandy
or stony portion of the soil. From the data thus
obtained an approximate estimate of the amount of
organic matter, sandy material, and clay can be made.
Next a test may be applied to see if chalk is present in
the soil in appreciable quantity. Add a few drops of
hydrochloric acid : if chalk is present effervescence
takes place.

Weather conditions and their effect on the land may
also receive attention and for this some apparatus will
be an advantage. For the most part the apparatus can
be home-made. It is a mistake to think that observation
of the weather necessitates special instruments. As a
matter of fact, simple home-made apparatus will probably give better educational results. There are different ways of making a "weather glass." A bottle
with water, into the neck of which is placed an inverted
glass flask, is a simple one. Another type is shown in
the examples of woodwork, p. 130. Note too the scarlet
pimpernel which is often spoken of as the shepherd's
weather glass. Weather lore, warnings, predictions, etc.,
that are proverbial may be used to give increased interest in the study of weather problems. Tests can be
made of the truth or otherwise over a lengthy period of
such sayings as, "A red sky in the morning, a shepherd's
warning" ; "A red sky at night, a shepherd's delight " ;

"Wet before seven, fine before eleven"; "Fog on the hill, water to the mill"; "A wind in the east, good for neither man nor beast." In fact it would be interesting work to collect the folk-lore of the district concerning the weather. By testing the truth or otherwise of weather sayings scholars will have. to study the weather at all times and not merely for a few minutes daily. Winds and their importance in connection with weather changes should also be observed and a wind vane should be made by the boys. A rain gauge is an instrument that can easily be fitted up at a very slight cost. A black bottle will answer as a collecting jar, while a tin funnel, costing about 2*d.*, will serve as a catchment area. A graduated measuring glass can be made from an empty medicine bottle.

Rain Gauge. Suppose the diameter of the funnel is found to be 4½ inches, then area of top of it will be

$$2\cdot25^2 \times 3\cdot1416 = 15\cdot89 \text{ square inches.}$$

The volume of water to cover this area to the depth of *one* inch would equal

$$15\cdot89 \text{ sq. ins.} \times 1 \text{ in.} = 15\cdot89 \text{ cu. ins.}$$

Suppose the area of the inside of the medicine bottle is found to be approximately 1·75 sq. ins., then

$$\frac{15\cdot89 \text{ cu. ins.}}{1\cdot75 \text{ sq. ins.}} = 9 \text{ ins.,}$$

which represents the height in the medicine bottle to which one inch of rain over the surface of the funnel would reach. Hence 9 inches in height of medicine bottle represents one inch of rain collected. For practical purposes the bottle can be graduated to measure half an inch of rain only. A strip of paper 4½ inches long is carefully divided into 50 equal parts and is then stuck to the outside of the bottle and a graduated measuring glass is at once available which will measure rainfall in $\frac{1}{100}$'s of an inch. In practice this rain gauge may not be absolutely correct but it will be sufficiently so for the purpose in view.

At one school I found that, in addition to an ordinary rain gauge, each boy had beside his garden plot a rain gauge of his own making and the rainfall had been calculated according to the following data.

Area of top of funnel of $2\frac{3}{8}$ ins. diameter

$$= (1\tfrac{3}{16} \text{ ins.})^2 \times 3\tfrac{1}{7}$$

$$= \frac{19}{16} \times \frac{19}{16} \times \frac{22}{7} = \frac{3971}{896}$$

$$= 4\cdot43 \text{ sq. ins.}$$

Now 1000 ozs. of water

$$= 1728 \text{ cu. ins.}$$

$\therefore$ 1 oz. of water $= 1\cdot728$ cu. ins.,

and since two tablespoonfuls of water weigh 1 oz., therefore one tablespoonful will measure $\dfrac{1\cdot728}{2} = \cdot864$ cu. ins.,

and since Depth $= \dfrac{\text{Vol.}}{\text{Area}}$, $\therefore \dfrac{\cdot864}{4\cdot43}$ equals rainfall in gauge with a funnel of $1\frac{3}{16}$ inch radius, and this is approximately **·2 inch.**

The rain collected is then measured in a medicine bottle graduated in tablespoons. The boys state what fraction of a spoonful and then take that part of ·2 inch. I am told that their errors as measured by the ordinary rain gauge are quite insignificant and it must be admitted that in measuring thus the boys are called upon to exercise a nice judgment at each operation.

Rain gauges are usually made with catchment areas
of either 5 or 8 inches diameter, so that if a tin funnel
of 5 inches can be obtained a measuring glass for this
standard size could be purchased.

Weather records as usually kept are not altogether
satisfactory. I have seen in some instances books full
of columns of figures concerning the temperature, rainfall, etc. Such columns of figures may look very nice
and imposing, but after all they are only figures. There
is no substance in them and instead of encouraging a
child to keep weather records on leaving school, they
would probably have the contrary effect. If the records
are shown as graphs a scholar must spend more thought
over his work, but even graphs after a time become
mechanical. Hence the teacher should try to vary
his work so as to prevent it becoming mechanical and
non-educational.

The plan might be tried of connecting the weather
records with the garden. For instance a graph might
be made of the growth of a broad bean growing out-of-
doors. Supposing that the plant for the first two weeks
made practically no upward growth, the scholars might
be called upon for reasons for this, and it would be found
probably that cold conditions prevailed. A few notes
descriptive of the weather conditions would then be
written on the graph. The next week the plant might
make considerably more growth and again the weather
conditions would be described by the scholars and these
written on the graph. The scholars would naturally be
keen to watch and think about the weather conditions
and that would be of greater value to them than many
columns of weather figures. Each scholar might have
one particular bean to measure. The rate of growth of

the bean plants would be found to depend upon the weather.   Exact measuring would be good work from an educational standpoint.   Each scholar should provide himself with a piece of stiff cardboard about 6 inches square which he should cut out as shown.   The cardboard strip should be put round the bottom of the bean plant, so that each measurement would be made from the same level surface.

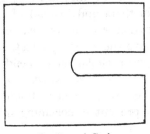

Cardboard Strip

A measuring stick would be required.   This might be a plain round stick about 5 feet high, with a piece of wire bent round with one end left as a pointer.   The wire should be so fixed that it could be moved up and down as required.   The pointer could thus be adjusted so as just to touch the top of the bean plant, noting that the arm is kept at right angles to the stick.   When quite correct a pencil mark could be put on the stick and the height of the bean measured therefrom.   I have known a broad bean plant grow a foot in height during a week of favourable weather (rain, warmth and sunshine).   The work in connection with a bean plant might last from January until June, and very marked contrasts would be noticeable.   For the first two or three months the side of a square might represent

$\frac{1}{10}''$ growth, while after this when growth is much more rapid let side of square represent *one inch*. Draw a line through chart separating the two scales.

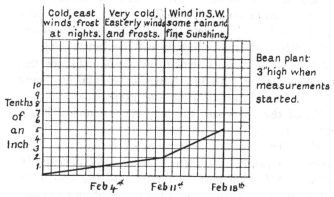

Graph of Broad Bean showing influence of weather on growth

In addition to keeping a graph of the growth of a plant, the temperature of the soil after changes of the weather should be taken. That is to say, supposing there is a week of cold weather, then the temperature of the soil at the surface should be taken and also the temperature at three or four inches below the surface. The readings should be repeated after a few warm days. Inferences made from comparisons of this kind and their effect upon germination and plant growth would be useful and tend to keep the weather records from getting into a groove. Another idea is to make a collection of the weather lore of the district and over a lengthy period test each saying as to its truth or otherwise. The amount of the local rainfall might be compared with that given in the newspaper for London. For instance the rainfall in London might be half that in one of the western counties and the cause of this difference should be ascertained. Matters of this kind can often be dealt with and reference made to such facts as rainfall of Lancashire and its influence on the cotton manufacture. Rainfall and heat and their combined effect upon plant growth can be noticed and dealt with, and then children will be better able to understand and appreciate many geographical facts such as those dealing with luxuriance of vegetation where there is abundance of rain and great heat ; or the aridity of the soil where the heat is great and moisture scarce, etc.

The formation and movement of clouds and their relation to rainfall and temperature may perhaps be noticed. Attention may be directed to the possibility of predicting weather changes by means of an examination of the sky. Dew, hoar frost, mist, fog, etc., and their formation and cause may be studied. Occasionally the class may be asked, after having been supplied with the morning's data of the observations of wind, temperature and pressure, to write down what the probable weather conditions for the day are likely to be. Such exercises will tend not only to make the scholars observers but will also train them to form conclusions from observed facts.

Where bee-keeping is possible a course of study dealing with bee life would be most useful and appropriate. A scheme of work based upon the following might be used. Familiarity with a hive and its contents. Life-history of the bee. Different varieties of bees kept. Preparation of the various parts of a hive for use. Swarming and how to deal with it. Feeding, supering, extracting, marketing, re-queening, increase.

I have thus shown that a very wide range of subjects more or less intimately connected with the garden can be utilised for nature study work. Perhaps as a conclusion to this chapter a few general remarks on nature study will not be amiss. Nature study covers a very wide and varied range of subjects, so wide in fact that its vastness often appals, and teachers think that before they can teach certain branches of the subject they must be walking encyclopaedias in them. Such need not be the case. Interest in the subject will soon overcome any difficulties. The teacher must often be prepared to say quite candidly to his scholars, "Together WE will try to find out what we can of this." Here then is one of the strong links of nature study work—the teacher and the taught may work from one level—the teacher does not lose in the estimation of his pupils as a consequence of confessing ignorance in a particular thing. On the other hand it is possible that it gives a child a better estimate both of his teacher and of himself by showing that even for a grown-up teacher, there are still things to learn—surely an excellent idea for children to have. For real nature study work teacher and taught must be keenly interested. As Professor Adams says, "Everyone who wishes to interest others must himself be interested."

6. THE GARDEN AND THE COOKERY CLASSES

IN the endeavours made to associate the school garden with other school subjects the question of its relation to the cookery classes might be considered. On the face of it there should be a most intimate connection. From the garden might be provided some of the vegetables required in the cooking lessons, while the cooking of some of the unusual vegetables, such as salsify, kohl rabi and others, should at times be undertaken by the girls, and the garden boys on those occasions might be allowed to visit the cooking classes and taste these vegetables and pass their opinions on them as regards flavour, etc. In this manner prejudices that exist concerning some of the uncommon vegetables might be broken down. The school herb bed should be made to supply what herbs are required and the girls might be allowed a small plot of ground on which to grow them. In after school days the herb bed is often under the care of the woman of the house. The drying of herbs for use in winter might also be undertaken. Where woodwork is combined with the garden many articles for use in the cooking classes might well be made by the boys, as for instance, jam stirrers, wooden spoons, wooden platform for bottling fruit, pot stands, etc. The garden might also supply small onions, shallots, red cabbage, cucumbers, etc. for the making of pickles by the girls.

The bottling of fruit is an important subject that should be taught in all cookery classes, and no great harm would be done if boys as well as girls could take part in the lessons. The subject is of great economic importance, for in years when fruit is plentiful, there is a tendency to wastefulness: moreover the home bottling of fruit can be highly commended as conducing to thrift and thought for the future.

Fruits can be bottled perfectly without special apparatus, a thermometer and patent bottles being the only requirements. The thermometer should be of glass with the figures large and clearly marked. Various patterns of bottles should be available and the advantages and disadvantages of each dealt with. Bottles with glass stoppers should be used as far as possible, so as to prevent possible contamination with metal covers.

When the bottles have been filled they should have the caps lightly screwed on and are then ready for sterilising. They can be placed directly upon a wooden platform standing in the bottom of an ordinary cooking pot, furnace, or other suitable vessel. There is absolutely no need to wrap the bottles round with hay, straw or cloths. The bottles may be allowed to touch each other. Water should be filled into the pot until it reaches up to the necks of the bottles. The sterilising should be done very slowly indeed, and when a temperature of about 160° F. has been reached, the pot may be removed from the fire, and if screw stoppered bottles are used, they should be taken out, one by one, and screwed tightly down. The bottles should then be put back into the hot water and allowed to stand in it until the next day, when they should be tested to see if hermetically sealed. The drying of fruit, particularly of plums and apples, might be attempted and hints for doing this are given with the drawing of a fruit drying tray, p. 145.

Other ways of developing an intimate association between the cookery, gardening classes and the school may suggest themselves to the teachers concerned. The cookery class should be an integral part of the school and not a thing apart. It would be a very beneficial thing for primary education if the Board of Education could do away with special subjects.

7. A SCHOOL GARDEN COMPANY

In the arithmetic lessons questions concerning stocks and shares often occur and it is a matter of difficulty to get children to realise that the value of a share may fluctuate according to circumstances. It was on an occasion of this kind that the thought struck me that a School Garden Company might afford a solution and at the same time assist in other directions. Accordingly I announced to the boys that we would form, from among the scholars and teachers, a School Garden Company whose capital should be split up into sixpenny shares. It was explained that the money thus provided would be used for purchasing a few extra tools and appliances for the garden as well as extra seeds. I further told the boys that each purchaser of a share would receive a share certificate and that if a holder of a share left school, his share might be sold to the highest bidder and that in any case I would personally guarantee that no boy should be a loser; that he should at least receive his share money back.

For income it was proposed to sell the produce raised on the Common Plot. As a result about 40 shares were taken up by scholars and teachers and, accordingly, I typed a number of share certificates which were in the following form:

.................COUNCIL SCHOOL GARDEN COMPANY.

SIXPENNY SHARE. No. ....

This is to certify that..........................is the Proprietor of One Share in the..........................Council School Garden Company, and that the said Proprietor is entitled to such share with the Profits and Advantages thereof.

As witness my hand, this.............day of..................One thousand nine hundred and eight.

Signed,

....
Head Master.

The bringing in of £. *s. d.* thus meant that a strict account of income and expenditure had to be kept, and this was done by the boys. This in itself taught a useful lesson of how accounts should be kept.

At the end of the first year's working it was found that we had bought a water-can, a garden basket, a syringe, foot rules for each garden scholar, string for garden reels for each boy, also seeds of perennials, kidney beans, stones for a rockery, and out of the profits 6*s.* was voted to provide each boy in the school

with an orange. A dividend of 2*d*. per share was likewise declared and paid. Just previous to the declara-
tion of the dividend, however, two of the holders of
shares left the school and their shares were offered
for sale and realised 7½*d*. each, while early in the
next year, after the dividend had been paid, a share
which was offered only fetched 6½*d*. During this year
other shares changed hands, mostly at a premium. No
more difficulty was experienced in dealing with share
sums, especially as the newspaper was consulted and
questions on stocks and shares therein quoted were
dealt with.

Before the end of the second year of the existence
of the Garden Company I gave up the charge of the
school and in consequence the Company came to an
end. Each share was paid back in full and 1½*d*. each
also paid as dividend.

Bee-keeping could in a similar manner be taken up
by means of a School Company, and in one instance
I know the fortunate possessors of shares in a School
Bee Company received 300 per cent. dividend one year.

8. THE RECORDING OF GARDEN WORK

What Books shall be kept for Garden Work?

What books shall be used for garden work and
in what form the records shall be kept are matters
which do not always receive the attention they merit.
Teachers are sometimes pressed to make their scholars
use what are termed rough note books. This has little

to recommend it. The notes are supposed to be written on the spot (in the garden) and at the moment. A boy engrossed in his practical work is very apt to forget, and rightly too, I think, his note book. If thought of, it has to be made up in the midst of his garden operations, irrespective of the state of his hands, etc. Boys, too, frequently work with their coats off, and the trouser pocket is no fit place for even a ROUGH note book when its owner is engaged in gardening. The boys do not realise what kind of notes are required or even how to set out what they have written. The writing, as may be expected, is scribbled horribly in pencil and the spelling is often bad. Such note books are not only useless but are positively harmful, inasmuch as the bad spelling and bad writing tend to become habitual. Note books can rarely have much time devoted to them by the teacher and, therefore, there is no correction of mistakes.

The notes where made usually contain the merest commonplaces, such as "planted potatoes, made a bonfire, pulled up weeds, etc."—information that would be better carried in the memory and produced when required in school. In after school days not one scholar in a thousand is likely to keep a rough note book in which to jot down observations, and the few who would do this would most likely have done it without preliminary note-taking in the garden.

The fact is that the keeping of rough notes requires much skill and discrimination. The kind of matter and the kind of observation required can only be decided upon when the scholar possesses knowledge of his subject, or the head teacher is himself a skilled observer.

The kind of descriptive work that can be undertaken has already been referred to in Chapter 4. Attention may therefore now be directed to where the writing, drawing, arithmetic shall be done. Unquestionably the ideal way is to do the composition in the ordinary writing exercise book as part of the composition work ; the drawing in the ordinary drawing book and so on. But school gardening is looked upon as a special subject and often someone is appointed to visit the school especially to see the garden work. When such is the case it is more convenient to adopt some method of keeping the records separate. How to do this satisfactorily is not an easy question. Some schools keep a separate book for each subject connected with gardening, in addition to the ordinary school books for these subjects. This is certainly a very confusing and wasteful method, for the books rarely get filled, especially where it is thought necessary to begin a new set of books each year. A nature study book with alternate leaves of writing and drawing paper is more satisfactory, and with few exceptions, e.g. graph work, all the records can be kept in the one book. The best method, however, is to do all the garden work on loose sheets of paper, written and arithmetic work on single sheets of ruled or unruled paper, drawing on loose sheets of drawing paper, graphs on loose sheets of squared paper, etc. The sizes of the sheets of paper used may vary and different coloured papers may also be employed. If cardboard modelling is done, then each scholar can make a portfolio—similar to the covers of a school register—in which to keep the loose sheets of paper. Covers could easily be made of some fairly stiff cardboard—simply two single sheets cut to the required size. The size of the covers should be a little larger than the largest sheet of paper the portfolio will contain.

GARDEN

RECORDS

John Smith

1912 – 13

Portfolio for garden papers

Covers of drawing books may also be utilised. If the ends of the string are twisted round and dipped in glue, then they will remain stiff for threading. The hole in the covers may be made with a bradawl if a punch is not available, and each piece of paper when completed should be likewise holed and threaded on the string. Each exercise when made should be dated. The exercises could be strung in order of date of working, or else drawings could be kept together, written work by itself, etc. This would be more easily accomplished if a leaf of coloured paper were placed between one set of papers and another. If a series of drawings, etc. were made these could be collected together. This method of keeping the garden records involves no special requisition for school stationery and no waste. If a scholar desired his garden book when he left school it would be in a handy and convenient form. A boy's gardening records for the whole time he is in the class, one, two or more years, might be kept together in this way and should show the progress he made during this time. By doing the work on single sheets of paper the scholar will be likely to make a fresh effort each time and do his best work, while when books are used, too often the pages do not improve in appearance as the work progresses. There is one disadvantage connected with working on loose sheets of paper to which it is well to refer. With the intention of making each boy's portfolio contain nothing but the best of work a master may occasionally let a scholar re-copy a badly written exercise, or not put in a bad piece of drawing. For the sake of the scholar a master should never allow this—all original work, good, bad or indifferent, should be duly strung in and if the boys see this they will soon try to do their best, while if second attempts are allowed then the work will never be the scholars' best.

It would be a useful piece of work for each scholar to make an index of what is contained in his portfolio, making the entry at the time the exercise is added. The method of keeping papers above described has been tried in several schools and found to be most successful.

Gardening note books, ruled in readiness for entering all sorts of facts, weather records, dates of sowing, thinning, transplanting, etc. can be bought. Intelligent teachers will give books of this nature a wide berth, for even if all the entries are completed of what conceivable use are the data? It may be argued that they

give dates for sowing, etc., but the seasons in this
country are rarely alike year after year. The time of
sowing, etc. depends not so much upon the time of
the year as upon the condition of the land—a cold and
wet spring might make several weeks' difference in the
date of sowing. Besides this, the information regarding
the time of sowing and harvesting most of the flower
and vegetable crops can be obtained in seed lists, garden
papers and other sources. Therefore it is better to
train the child to know where to seek information in
case of doubt and difficulty than to waste his time
making useless entries. Whatever method of keeping
notes and records is adopted, it must not be expected
that all the scholars can reach the same standard.
From many points of view the work may be dis-
appointing to the teacher, which is far from being a
bad sign, if it calls forth efforts to raise the standard
of work. On no account should the written work be
copied from the blackboard. One other point; do not
give "a lesson" on the subject of the essay just previous
to the writing thereof, or the work produced will not
be that of the scholars—a fact that may easily be
determined after reading a few of the contributions!
Mark the work as soon as possible after it is done and
then, if the necessity arise, give the lesson, for you
will have found out the children's knowledge and it
should be an easy matter to correct faults and make
good omissions. Mistakes in spelling should be corrected
by scholars at the foot of the composition. Expect a
high standard of work and you will make your scholars
aim high.

9. THE SCHOOL MUSEUM

A SCHOOL museum may form a valuable adjunct
to a school. It should contain examples of small
animal life, etc. which would be available for examination and comparison at times. The specimens, too,
would be used in the composition, drawing, and other
lessons. The school museum is so often a heterogeneous
collection of rarely used things that there is real need
for giving a few hints concerning what may be done.
Of course it is not suggested that the museum should
be restricted to the type of specimens here described.

In any garden there are large numbers of small
creatures that will well repay study, and one or two
specimens of each can be collected and preserved for
use when occasion requires. It is a matter of surprise
what a collection can be built up when once started.
In work of this kind boys are exceedingly keen and
interested. Grubs, slugs and other "fleshy" specimens
are best kept in a preservative fluid in glass tubes
which are then corked. The specimens must not at
first be put into the full strength spirit or they will
be likely to harden, shrivel up somewhat, and become
more or less discoloured, thus presenting anything but
a natural appearance. To get over this difficulty I
proceeded as follows :—The creature to be preserved
was put for a few hours into a mixture of water and
methylated spirit, using about half and half of each. It
was then taken out and as the legs, etc. were still soft
enough to be moved they were arranged in as natural
a position as possible. This done the specimen was put

into a solution containing about 75 per cent. of methylated spirit to 25 of water. It was left in this for about a day and then transferred to a vessel containing full strength methylated spirit in which it was allowed to stay for a few days. Larger specimens were given a longer time than smaller ones. If the specimen discoloured the methylated spirit it was taken out and put into a fresh lot. The specimen was now ready for bottling for the museum. For this purpose I used a weak solution of mercuric chloride in methylated spirit. A chemist will make up this if told for what purpose it is to be used. It is very poisonous and therefore must be kept under lock and key and away from the scholars.

Do not fill the tubes quite full with the fluid and cork with good sound corks that have been well pressed. See too that the corks do not touch the fluid. Grubs such as those of the cockchafer beetle often turn black when kept in spirit. This can be prevented to some extent by dropping the grubs into boiling water for a minute or so before putting them into the spirit as above described. In the case of large specimens, snakes, frogs, toads, etc., I have driven the corks in a little below the top of the bottle and then covered them with plaster of Paris which is afterwards painted over with Brunswick black.

Sometimes with all this trouble it may be found that a specimen has discoloured the liquid in which it is preserved. If so, withdraw the cork, take out the specimen and put it into some methylated spirit for an hour or so and then put it back into a fresh tube with new preservative fluid. The advantage of using methylated spirit instead of pure alcohol is that it is less expensive. I have found the above method entirely

successful even in the preservation of fully grown frogs,
toads, and a snake (2 ft. 8 ins. in length). Frogs' eggs
before being put into the preservative were first put
into boiling water.

The tubes when completed are put on stands and
name, date, and other particulars written on. The
stand may be made for a single specimen or for a series
showing the life stages. The wooden stand is so made
that there is room to display the perfect insect in the
case of flies, butterflies, etc. in front of the tube (see
p. 108).

It will be necessary to stock glass tubes in various
sizes. They should have flat bottoms. They can be
obtained through the usual educational supply companies
or, if any difficulty is experienced, application can be
made to Messrs Flatters and Garnett, Manchester.
Spiders are best preserved in turpentine. Instead of
using the fluid above described a strong solution of
formalin in water may be used.

Another method of treating larvæ is to remove the
inside and then inflate and dry. This process is more
difficult than that already described but it has advantages
inasmuch as the specimens can be seen without any
distortion, as sometimes occurs when specimens in
spirit in round tubes are viewed. Another advantage
is that specimens may be mounted in natural surroundings by wiring them to twigs, a piece of bark, etc.
Thus set up they are very attractive. The larva should
first be killed by immersion in methylated spirit. A
small vertical hole is then cut in the anus and the
specimen laid between a few sheets of blotting paper
and the contents of the body squeezed out. Start
pressing near the head first and gradually work down

the body. After completely emptying the skin the tube of a small blow-pipe must be inserted in the opening and the skin tightly tied round it with fine silk. The skin is gradually inflated and dried as quickly as possible before a fire or over a spirit lamp. Continue the inflation through the blow-pipe until the specimen is perfectly dry. Untie the larva from the blow-pipe and gum the pro-legs to a twig or a card, etc. or wire it to twig, bark, etc. In some instances after a larva skin has been emptied fine sand is inserted through a tiny funnel. It is hardly necessary to remark that this process of preservation requires some practice to do nicely. I have had scholars who were fairly successful in the blowing of larvæ.

The perfect insect has next to be considered and in most cases this is not a matter of great difficulty. For most specimens a killing bottle is necessary. A jam-jar half filled with bruised laurel leaves and kept tightly corked can be used, but a better and more effective one can be made for a few pence. Procure a stout one pound size jam-jar to which a sound cork has been fitted, and take to a chemist and ask him to make a "killing bottle." It is also advisable to have three or four mounting boards of various sizes, while a small box of assorted sized setting pins and a pair of forceps would be almost a necessity for doing good work. Most insects are set with the wings fully expanded but it is a good plan to mount more than one specimen of each kind. Thus one could be set in the usual way with wings expanded and the other as it appears in an attitude of rest. In many cases the employment of this latter method will allow of the specimen being mounted on a piece of bark, etc. and

thus can be shown how its colour matches the surroundings.

Beetles may be mounted, one specimen with wings
expanded and another with the wing cases closed.   In
the case of beetles a setting pin of suitable size is
selected and this is put through the right-hand wing
case, a little way from the joint of thorax and abdomen,
thus:

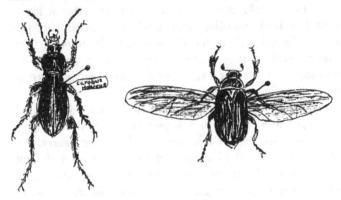

Beetles Pinned

It is necessary to arrange the legs, etc. as naturally
as possible and until they get set these may have to
be held in position by means of ordinary pins.   A slip
of paper about an inch long and from an eighth to
a quarter of an inch wide may be fixed on the pin
below the body of the beetle.   On this would be written
name, date and any other desirable particulars.

A fly or a butterfly has a suitable setting pin thrust
through the middle of the thorax and its body placed
in the groove in the mounting board.   The wings are
then gradually drawn out, beginning with one of the
fore wings.   A needle or pin very carefully handled

will answer for this purpose—placing the pin as near
a strong wing vein as possible. When the wing is in
position a narrow strip of paper is pinned over it and
held in position by means of ordinary pins. In a similar
manner the other wings are carefully arranged and held
in position by strips of paper. Instead of strips of
paper, bristles from a bass broom are sometimes employed. The specimens are left on the boards for
several days until they are dry, when they can be
transferred to a specimen case or direct into the
museum. They may be named in the same manner as
described for beetles.

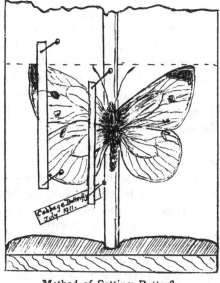

Method of Setting Butterfly

Moths often require to have the contents of their
abdomens removed. The abdomen is cut off and the

inside can then be taken out by means of a bent pin, taking great care in handling the part, that the scales are not rubbed off. I usually wrap the part in cotton wool when handling it. When the viscera are removed, tiny plugs of cotton wool which have first been rubbed on some arsenical soap are packed into the abdomen so as to make the body retain its natural shape and appearance. When the operation is completed, the part should be stuck back in position by means of an adhesive.

Scholars with very little practice will become quite adepts in setting insects and the work should, as far as possible, be done by them. Thus the museum becomes part of their making and that means something. Naphthalene, camphor, or albo-carbon should be put with the specimens to ward off mites. Specimens will sometimes turn mouldy and may then be cleaned by means of sulphuric ether, applied by means of a camel-hair brush, just dropping the liquid on the mould. If this application does not remove it, then the brush itself must be used, but very gently.

Some means of classification should be adopted. Those creatures that are useful in the garden, such as lacewing fly, hover fly, ladybird, ichneumon flies and others may be kept together, while another group subdivided according to the plant they infest may be made of injurious creatures. Another plan would be to group the injurious and useful creatures connected with some plant food into a group, e.g. all the injurious and useful creatures that are found on the cabbage family of plants.

A collection of snails can easily be prepared for the museum by putting them into boiling water for

a minute and then removing the bodies by means of
a bent pin. The shells can afterwards be mounted
on cardboard strips.

Those who are keenly interested in the preservation
of specimens of small animal life and wish to display
them to better advantage and with less risk of damage
than by leaving them loosely on a shelf of the museum
could make cases such as described on p. 135.

A collection such as indicated is not difficult to
make. It is quite inexpensive. The work of mounting
and preparing the specimens can be undertaken by the
scholars who thereby must use "hand and eye" with a
nicety. Thus the work is educative and is likely to
provide hobbies in later life.

10. EXPERIMENTAL WORK IN THE GARDEN

A MOST important phase of school gardening is that
connected with experimental work, and yet it is a part
of the subject too often overlooked, or only scantily
touched upon. The working of simple experiments in
the garden dealing with plants and plant-growth tends to
give the scholars an "attitude of mind" that will in after
life be of the utmost service to them. Amongst other
things it should show that there are more ways than one
of doing a certain thing, and that there are reasons why
certain methods are better than others. It should show,
too, that if success does not come one way other ways
may be tried. Experimental work goes down to the
reason of things, the "why" and "wherefore" of them.
The work has a very stimulating effect upon the minds
of the scholars.

The experiments selected should be such that the results are forthcoming in from one to three years, the length of time a scholar is usually in the class. In most instances the experiments should show decided results and it is better if a series of observations can be made in connection with an experiment, as that means continuous interest. It is of the highest importance that careful and complete records of all experiments conducted should be made, noting the conditions prevailing and results achieved. Special attention should be given to this part of the work.

Artificial manures are not recommended for experiment in the school garden unless the teacher knows something about them. One or two light dressings of nitrate of soda might, however, be applied to a portion of, say, a spring cabbage bed. Some of the cabbage plants might be supplied with rather liberal dressings to show the strength of the substance and its ill-effects in large quantities upon plant life.

There are, however, numbers of simple experiments on plant-growth that will add materially to the pleasure and interest of the school garden. The following was worked by my boys with broad beans. Early in 1909 the top class in school was given about two quarts of broad bean seed and told to select about 200 of the finest, biggest and heaviest beans. The scholars used their eyes, picked out beans, balanced them in their hands, and then having made up their minds selected the beans or rejected them. This work was done in an arithmetic lesson and in some quarters it might have been looked upon as out of place, but it required very nice estimation, discrimination and judgment on the part of the scholars to make the necessary selection and,

that being so, the work was educative. When the beans
had been picked out the chemical balance was brought
into use and each boy in turn weighed two or three
beans, the weights being recorded. The 60 heaviest
beans were finally chosen and their weights in grams
are shown below :

	(a)	(b)	(c)	(d)	(e)	(f)	Totals
1	·262	·242	·228	·220	·214	·209	1·375
2	·252	·242	·227	·219	·213	·207	1·360
3	·251	·240	·226	·218	·213	·206	1·354
4	·251	·239	·226	·217	·212	·205	1·350
5	·250	·235	·222	·217	·212	·205	1·341
6	·249	·232	·222	·217	·211	·205	1·336
7	·248	·230	·222	·217	·211	·204	1·332
8	·248	·229	·222	·216	·210	·203	1·328
9	·247	·229	·222	·215	·209	·201	1·323
10	·246	·228	·220	·215	·209	·200	1·318
Totals	2·504	2·346	2·237	2·171	2·114	2·045	13·417

Total weight = 13·417 gms.
Average weight per bean ·224 gm.

In a similar manner another 200 of the very smallest
beans were sorted out and afterwards weighed, the 60
smallest giving these results :

	(a)	(b)	(c)	(d)	(e)	(f)	Totals
1	·086	·093	·104	·114	·124	·118	·639
2	·093	·121	·097	·116	·124	·107	·658
3	·069	·097	·111	·109	·132	·128	·646
4	·106	·091	·103	·118	·118	·126	·662
5	·111	·101	·119	·096	·113	·100	·640
6	·085	·102	·112	·103	·115	·116	·633
7	·078	·109	·106	·124	·137	·137	·691
8	·076	·123	·125	·111	·115	·124	·674
9	·102	·090	·098	·130	·109	·132	·661
10	·078	·106	·118	·132	·129	·133	·696
Totals	·884	1·033	1·093	1·153	1·216	1·221	6·600

Total weight = 6·600 gms.
Average weight per bean ·110 gm.

B.

As already stated each boy took a share in the weighing, and it furnished a splendid lesson in accuracy. The setting down of the figures, totalling the rows and testing by cross adding afforded good practice in addition. The beans were planted in four rows as shown :

First row.   Single row of 20 heaviest beans.
Second row.   ,,   ,,   ,, 20 lightest beans.
Third row.   Double ,,   ,, 40 heaviest beans.
Fourth row.   ,,   ,,   ,, 40 lightest beans.

The ground had received the same treatment throughout and the rows were of equal length. The boys were not told what results to expect, although they naturally ventured opinions. They were advised to note any differences in growth in the rows. These soon manifested themselves, for the beans from the largest seed outgrew the others and made stronger-looking and healthier plants than those from the smallest seed. Unfortunately during the Whitsuntide holiday the beans became badly infested with "black fly" which somewhat spoilt the crop. In due course the beans were harvested and when thoroughly dried were weighed with these results :

		lb.	oz.
First row.	Single row, heaviest beans.	1	7
Second row.	,, ,, lightest beans.	1	4
Third row.	Double row, heaviest beans.	1	13
Fourth row.	,, ,, lightest beans.	1	1

Thus the boys were able to see that the best and finest seed gave the best results. The question whether single or double rows were the better was discussed and it was agreed that the evidence was not conclusive enough and the boys expressed a wish to test further this point in the next year. The beans from each of

the four rows were sorted out and duly weighed. From
the first row the 20 heaviest beans were :

·235 ; ·218 ; ·239 ; ·233 ; ·210 ; ·224 ; ·216 ; ·232 ; ·242 ;
·212 ; ·224 ; ·229 ; ·227 ; ·222 ; ·208 ; ·220 ; ·197 ; ·248 ;
·224 ; ·232.

Total weight 4·492 gms., giving an average weight of
·225 gm. This was against an average weight of bean
planted of ·223 gm. The 20 heaviest beans from the
second row weighed 2·773 gms., giving an average of
·138 gm. The third row gave these weighings for the
10 heaviest beans ·273 ; ·270 ; ·276 ; ·251 ; ·302 ; ·237 ;
·255 ; ·251 ; ·285 ; ·243. Total weight 2·643 gms., average
·264 gm. This result was an improvement on the seed
planted in respect of weight of individual beans and
also in average weight. The next 10 heaviest beans
weighed 2·372 gms. Thus the 20 heaviest beans from
this row gave a total weight of 5·015 gms. and an average
weight of ·250 gm. The fourth row gave 3·223 gms.
as the weight of the 20 heaviest beans, an average
of ·161 gm. The two heaviest beans from this row
together weighed ·456 gm.

In 1910 the experiment was continued along the
same lines and seed was planted as follows :

First row. A double row of the 60 heaviest beans
selected from seed of 1909 experiment. Total weight
13·58 gms. Average weight per bean ·226 gm.

Second row. A double row of the 60 smallest beans
selected from seed of 1909 experiment. Total weight
9·347 gms. Average weight per bean ·156 gm.

Third row. A single row of the 30 heaviest beans
selected from seed of 1909 experiment. Total weight
6·957 gms. Average weight per bean ·232 gm.

Fourth row. A double row of 60 new seeds obtained from seedsman 1910. Non-selected. Total weight 11·730 gms. Average weight per bean ·195 gm.

Fifth row. A single row of 30 beans, new seed obtained from seedsman 1910. Non-selected. Total weight 5·570 gms. Average weight per bean ·186 gm.

During the period of growth the boys frequently measured the height of the beans and interest in the experiment was thus maintained. The heights on three dates were as follows :

	May 26th.	May 30th.	June 7th.
Row 1.	19·5 ins.	24 ins.	34 ins.
,, 2.	15 ,,	20 ,,	30 ,,
,, 3.	17 ,,	22 ,,	31 ,,
,, 4.	14·5 ,,	18 ,,	27 ,,
,, 5.	13·75 ,,	17·5 ,,	25 ,,

A graph of the growth of each row of beans with weekly measurings could have been made and each boy might have had one particular bean in each row to record graphically. On the 12th September the beans from the experimental rows were picked and the seed weighed as follows :

					lb.	oz.
Row 1.	Double row of	60	heaviest beans	gave	5	3
,, 2.	,,	60	smallest beans	,,	3	6
,, 3.	Single ,,	30	heaviest beans	,,	4	5
,, 4.	Double ,,	60	non-selected beans	,,	3	7
,, 5.	Single ,,	30	,, ,,	,,	2	15

The length of each row planted was the same throughout and other conditions were also alike. The experiment was one which the boys who were only in the gardening class one year were thoroughly able to appreciate, while those who had taken part in it for two years became enthusiastic believers in selected seed.

The experiment called forth work in selecting the beans, weighing these accurately, noting the rate of growth of the beans and measuring the same. Graphs might have been made to illustrate the growth, and weight of produce. Various arithmetical problems relating to the experiment were also solved, as, for example, finding percentages of increases, etc. ; value of an acre of land planted with beans in (*a*) double rows, (*b*) single rows ; the cost of the seed in each case to be reckoned. Comparisons between the crops of 1909 and 1910 were made and the ill-effect of black fly on the crop noticed. Various composition exercises relative to the experiments were also written. The value of selected over non-selected seed was well brought out and thus the boys saw in this experiment that the finest and best seeds produced the best results, and therefore the need for selection of seed.

Another experiment of the same nature might be made by saving pods of peas, broad beans and runner beans for seed and when ripe shelling them into groups according to the number of seed each pod contained. Thus the peas might contain one, two, three, four peas and so on up to nine or more peas per pod. The seed from all pods with four peas in would be kept in one heap, the pods with six peas in would be kept in another heap, etc. Similarly broad beans would be shelled into heaps according to the number of beans the pods contained, and runner beans would be treated in the same way. If when this had been done the seed in each case was sorted out and only the finest specimens chosen, then the selection would have been carried a stage further than had been done in the experiments with broad beans, for it is possible that fine and large seed may not always mean fine, long and well-filled pods. The seed from pods of nine peas might be planted in one row and an equal number of peas from pods of two planted in a second row. Broad beans and runner beans might be treated similarly.

Other experiments suitable for school garden work may now be indicated. Winter greens and brussels sprouts do better as a rule when planted in ground made firm, and this can be shown by planting a row of brussels sprouts and a row of curly kale in ground made firm, and also planting one row of each in ground that has not been made firm. The crop results should be noted.

The need for deep digging and proper manuring for such crops as carrots, etc. can easily be shown experimentally thus: Divide a plot of ground lengthways into three strips of equal size. Dig the first strip deeply and manure in the autumn. Dig the second strip ordinarily in spring and add manure. Dig the third strip of ground ordinarily in spring but do not add manure. Sow carrot seed in rows across the three strips and note results.

Parsnips for exhibition purposes are sometimes planted thus: A hole of four to five inches diameter and about two feet deep is made with a crowbar or similar tool. The hole is then filled with carefully prepared soil and firmed. On top of each hole are sown two to three parsnip seeds which are later thinned out to one plant for each hole. As an experiment one row of parsnip seed might be planted according to this plan, another row being planted in ground that has been double dug, and a third row in ground dug ordinarily, and the results noted. The effect of thinning could be shown by growing two rows each of such vegetables as carrots, beet, parsnips. Thin out one row of each and leave the other row unthinned. At the end of the season compare the crops.

Seed potatoes that have chits on might be sorted into two groups, one containing potatoes with only one strongly developed shoot, the other with potatoes with two or more chits on. Plant in separate rows to see if there is any difference in the crops.

Onion culture might provide an interesting and useful experiment conducted as follows :—Divide a packet of onion seed of such a variety as Ailsa Craig into four equal parts and plant in this way :

First portion sow out of doors in the autumn—the 2nd or 3rd week in August.

Second portion raise in heat in January.

Third portion raise in a cold frame or school window —plant in January.

Fourth portion sow out of doors at ordinary time.

The autumn sown onions should be transplanted in March and the onions from the second and third lots should be planted out about the second week in April. The advantages and disadvantages of each method of culture should be noted. The crops might be weighed so as to give accurate returns.

The keeping qualities of the onions raised by each method could also be ascertained.

Further experiments with onions dealing with cultivation could be carried out as follows :—Divide a bed of young onions into seven equal parts :

(*a*) The first portion to be hand weeded but not hoed.

(*b*)   This portion to be hoed but not hand weeded.

(*c*)   This portion to be hand weeded and dutch-hoed once a fortnight.

(*d*)   Treat this portion as (*c*) except that it is to be dutch-hoed once a week.

(*e*)   Treat as (*d*) but in addition apply a little weak manure water once a month.

(*f*)   As (*e*) but manure water applied weekly.

(*g*)   Treat as (*d*) and apply a little weak manure water one week and the next week a very light dressing of mixed artificial manure.

In some gardens the same land is cropped with onions for more than one year.   Whether this is advisable or not can be determined by growing one-half of a crop of onions on the onion bed of the previous year, while the other half could be planted in ground that had not borne onions.   Of course both the new bed and the old bed should be deeply dug and liberally manured, the plots being treated alike in these respects.

The effect of deep digging could be shown by having a strip of ground divided lengthwise into three strips of equal width.

> One strip to be double dug.
> One   „   „   „   dug ordinary.
> One   „   „   „   „   shallow.

Plant across the patch onions, carrots, potatoes, and note results.

Coal cinders and ashes are in many localities, especially where the soil is heavy, often put on the gardens, sometimes in fairly liberal quantities.   Opinions differ much as to whether the practice is to be commended or otherwise. Whether or not coal ashes are harmful can be shown by means of an experiment, in fact the experiment should be continued for two or more years on the same piece of ground, adding fresh ashes each year. While a dressing of coal ashes might improve the texture of a very heavy soil, it is the continual use of the substance on the same ground that needs investigating. Mark off a plot of ground into three strips lengthways and treat the first strip with coal ashes, the second strip leave as it is, while the third strip could be supplied with ordinary manure. Plant across the strips crops of carrots, onions, potatoes, cabbages, etc. In due course the crops should be examined, particularly as to quality and freedom from blemish.

It is often noticeable that the potatoes at the ends of a row are the finest plants and yield the best crops, probably owing to the advantage they enjoy of having more light and air than the other plants can obtain. Hence it might be well to plant some potatoes at wider distances apart than usual and compare the crop results with some of the same kind of potatoes planted the usual distance apart. From the data obtained from the experiment calculations might be made of cost of seed by each method required for an acre of ground and also the value of crop over the same area.

The culture of out-of-door tomatoes is very popular and a few simple experiments could easily be undertaken with these plants.

(*a*) Put tomato plants out of doors in ordinary soil and a like number of plants in ordinary soil to which a little lime or old lime mortar has been added and note effects upon the growth and productiveness of the plants.

(*b*) An out-of-doors tomato crop might have one portion kept well watered and the other portion given no water and the effect of the treatment in each case noted.

(*c*) Pick out side shoots as they develop from one or more tomato plants and allow one or more plants to develop their side shoots. Compare earliness of crop and quantity produced in each case.

Double cropping, as it is termed, might be given trials in the school garden. An experiment to ascertain its value could be conducted by planting a few rows of early potatoes; keeping the rows about six inches wider apart than usual. Between the rows plant another crop such as brussels sprouts. Find the value of crops from potatoes and sprouts. An extension of this experiment could be made by planting a similar sized piece of ground with early potatoes, lifting the same as early as possible and then planting the patch with brussels sprouts plants obtained from the same source as those used in the preceding experiment. The crop results should be noted and comparisons made between the two methods of cropping. Further work of this nature could be undertaken by planting broad beans with spring cabbage; broad beans with potatoes; radish on an asparagus bed; potatoes and winter greens; spring onions or lettuce between the rows of strawberries; lettuce on the sides of the celery trenches; parsley and other herbs as well as mustard, cress, lettuce and radish around the edges of the plot. Such experimental work as this should lead children to be always on the alert to utilise the ground they cultivate to its fullest extent.

Lettuce and green crops might be planted in drills or shallow trenches and the effect of watering in such circumstances might be compared and contrasted with similar crops not planted in trenches. Another portion of the crop might be planted in the ordinary way and kept well dutch-hoed.

Ground well covered with chickweed might have this crop dug in for manure. In some instances, too, mustard might be grown and dug in as a green manure. This work should show the value of "green manuring" and the scholars should be made to realise that crops thus dug in give back to the land all the food they have taken from the soil, plus the food obtained from the air. In the case of digging in mustard there is also a certain effect upon the pests in the soil.

A few Paradise stocks might be planted in the fruit plot and the next year the surface of the ground around these well manured to encourage growth of fibrous roots. When ready for grafting take up two or three of the stocks, graft them indoors and then replant. Graft some of the undisturbed stocks and the results in each case can be compared. (It should show the influence of roots on the future of the trees. The stocks lifted would naturally have their roots damaged more or less and thus would have double work to perform.) Paradise stocks can easily be obtained by layering shoots.

The effect of mulching gooseberry and currant bushes might be tried and also the mulching of such crops as peas, beans, etc.

Various ways of getting rid of slugs might be tried.

(*a*) Place a piece of brown paper on the ground near a crop infested with slugs. The paper may be kept in position by placing a stone on each corner of it. Slugs will hide under the paper and may then be taken and destroyed. (*b*) Grease a cabbage leaf and place beside the brown paper trap. (*c*) Try rings of soot, lime or sawdust round individual plants to prevent attack. (*d*) Try sawdust on which has been sprinkled a little carbolic acid, or other disinfectant, as a barrier to slugs. (*e*) Place a little powdered alum round a few plants to keep slugs away.

The habits of earwigs might be discussed and various forms of traps suggested by the boys for catching them. A match-box with moss in it and the lid left a little open will make a first-rate hiding-place for these creatures if put in places where they congregate. They may then be taken and destroyed.

Different methods of planting early peas might be tried ; a variety such as the Pilot would be suitable for this purpose. Divide a packet of seed into three lots. Plant one portion in November, another in February, and the remaining lot in March. Note results in each case as regards (*a*) earliness of picking, (*b*) condition of crop, (*c*) weight of crop and value.

The value of sticking can be demonstrated by growing two rows of peas (e.g. Gradus), sticking one but not the other. Note the difference in the labour of picking.

Place in a box two or three diseased potatoes with a few sound ones of different varieties and note which, if any, of the sound potatoes become affected with the disease. Following up this idea reference may be made to the very bad practice of leaving small and diseased potatoes on the land after the crop has been raised and thus infecting the ground. This matter might be tested thus: on a portion of ground which has not carried a potato crop for one or more years, put a few diseased potatoes and leave them on the ground throughout the winter. In the spring divide this plot of ground into two parts, giving one of them an ample dressing of lime. Plant potatoes on the two parts and also plant some on another piece of ground that has not carried potatoes for two or three years. Note results in each case.

The question of using "old" seed is one that often arises in a small garden. The matter may be tested by taking some seed of two or three kinds which is, say, two or more years old. Plant side by side one row of the old seed and one row of new seed. The time taken for the old and the new seed to germinate, and the after vitality and crop should be taken into account.

The same type of experiment as described for broad and kidney beans, in connection with selection of seed, might be tried with potatoes. In a plot of early potatoes mark those plants which look the most robust and sturdy—not necessarily those carrying the greatest quantity of haulm. Dig these selected roots in due course and if crop is satisfactory in each case save the potatoes for seed. Likewise from the same plot put by some potatoes obtained without any selection from the other roots. Treat the seed potatoes alike in each case and the next year the selected potatoes may be tried against the non-selected ones.

While dealing with potatoes there is another matter worth consideration, namely, the degree of ripeness or otherwise of the potatoes when put aside for seed purposes. It has been said that Scotch and Irish seed potatoes are superior to English ones, because in those countries the tubers for seed are not ripened so much as in England. Here then is subject for investigation which can thus be carried out. From a growing crop of early potatoes, the seed of which was, preferably, obtained from a distance, make the following selections.

(*a*) Four of the finest and most healthy-looking roots. Dig these while the skin can easily be removed from the tubers with the finger. Save all the potatoes except the very small ones for seed. Leave these potatoes on the ground for two or three hours and then put them into shallow boxes, with eyes uppermost. Use new boxes for this experiment so that there is no fear of contamination through diseased potatoes having previously been in the boxes. Place the boxes and potatoes in a light room and keep thus until planting time the following year. Take care of course to protect from frost.

(*b*) Another four of the finest and best potato plants should be left thoroughly to ripen their tubers. Again save all except the small for seed. Treat exactly as in the case of (*a*).

(*c*) Take unripe potatoes that have not been selected and treat exactly as in the case of (*a*).

(*d*) Take ripened potatoes similarly and treat as (*a*).

The experiment, if desired, could be extended by testing the different methods of storing potatoes for seed purposes, e.g.

(1) Save unripe seed from selected plants and store in shallow boxes, as was done in (*a*) above described.

(2) Select unripe seed and store until spring in boxes, bags or other fashion.

(3) and (4) Treat ripe seed similarly to (1) and (2).

An experiment of this nature could be conducted for two or more years and thus its conclusions should be more definite.

Numerous other experiments could be tried with potatoes dealing with sizes of seed tubers, whether advisable or not to cut tubers of early or main crop potatoes for seed purposes. Rows might be moulded or earthed up and others left without earthing up and results noted.

Leeks may be planted (*a*) in trenches and treated in the same way as celery ; (*b*) on the surface of the ground and afterwards earthed up; (*c*) in a row of holes made with a dibber or other suitable appliance. The holes should be about 10 to 12 inches deep and about six inches apart. Into each hole drop a leek, keeping the plant in an upright position. Drop a little soil in on the roots. As the leeks grow in the holes they will become bleached, for rain, wind and other agencies will gradually cause the holes to become filled with soil. Apply liquid manure occasionally to a portion of each of three lots of leeks, particularly noticing the ease with which the manure was applied direct to the leeks grown in holes and the little waste in its application. When judging the crop results take into consideration the labour connected with each method of culture.

The effect of flowers of fruit trees and bushes being visited and fertilised by bees, etc. may be tested. Two or three gooseberry and currant bushes may have muslin bags tied over unopened trusses of blossom. It will probably be found that the blossoms protected by the bags will set practically no fruit.

In school simple experiments might be conducted dealing with the germinating power of seeds. The seeds to be tested can be placed on blotting paper on a plate. Water may be added to the plate from time to time so as to keep the blotting paper moist. In other cases boxes of moist sand or soil might be used. The number of seeds used in any test should be counted and thus the percentage of seeds that germinate can be found and hence the relative values of the samples of seeds.

Too much in the way of experimental work should not be undertaken at one time or the very plenteousness of it may confuse. Get the scholars to enter into the spirit of the experiments and their work and character will improve in the school as well as in the garden. Naturally some of the experiments would require considerable investigation and a large number of cases need to be examined before a general rule or law can be assumed or established. However, experimental work will naturally give scholars food for thought and that alone is extremely beneficial to them. But scholars will not only gain knowledge through experiments, they are likely to tell their fathers of their experiences and so help to effect much-needed changes in the home gardens.

11. GARDEN HINTS

THERE are certain facts concerning the school garden, commonly overlooked, that are worth notice and consideration. Each plot should be marked out by strong pegs driven in at the corners and then the edges of the plots can be kept straight and nicely edged up with little trouble by stretching a line from peg to peg and cutting the ground with a spade, which should for this purpose be used with the back of the blade turned to the front. The care of the paths tends to give an orderly look to a garden. The whole of the garden land should be so laid out that as much as possible of it is closely cropped. The idea of getting the maximum amount of produce in the shortest time should be the aim throughout. By this is meant not only getting the largest yield from any given crop but also raising as many crops as possible from the land each year. Too often allotments and gardens can be seen only partially cropped for extensive periods of the year. Intensive cultivation can easily be arranged in the school garden. Some crops may be planted earlier than usual. Broad beans are frequently not planted until March and the crop is not ready for gathering until July or later. The beans, too, during this period are likely to suffer from " black fly " and to some extent also from dry weather. Now if the beans are planted as early as possible in January, then the crop would be ready for gathering in May or quite early in June, a time of course when the beans would be worth most. But another and greater advantage from the earlier planting would be the opportunity afforded of clearing the ground and planting a second crop such as brussels sprouts, turnips, or other vegetables. Brussels sprouts thus planted out in June would probably be much more profitable than if they followed beans cleared off the ground in July. In September a row or two of winter lettuce may be sown on fairly firm ground and in early spring the young plants that have come through the winter could be planted between rows of peas, beans, etc. Thus treated they should be fit for use in May, a time of the year when lettuce is scarce and dear. The ground too is carrying two crops at the same time, peas and lettuce. Lettuce is often planted in rows across the plots in similar manner to other crops, but unless they are so planted as a "catch crop" ground is wasted. There is always plenty of room along the edges of the plots for crops such as lettuce, radish, parsley without in the least interfering with the other crops. Lettuce should be grown continuously through the year, a little seed being sown at intervals and transplanted when occasion may arise. Thus treated, land usually given to lettuce might be used for some other vegetable. The point of course is to make the scholars realise to the fullest extent the comparative value of the crops and also that by a little foresight the land can be made to yield more. If there are any unoccupied corners of the garden, think of some crop that can be grown thereon, or some other way of utilising the ground. A marrow plant, flowers, a few cuttings of fruit bushes, etc. may be found to answer for this purpose. The paths between the plots should be made narrow, a foot wide is quite sufficient, as only walking room along them is required. The main pathway should be two and a half to three feet in width.

When there is a hedge around the land, if it is yours,
keep it trimmed, but in any case keep down the
nettles, docks, etc. usually found growing at its base.
In some instances the hedge bank may be utilised
for growing vegetable marrows, ridge cucumbers or
rhubarb. Every part of the garden should be kept
neat and trim, as neatness and orderliness cannot be
too much impressed upon the attention of scholars by
means of good examples. In some gardens there is
often an unsightly rubbish heap which could be screened
from view for at least a large part of the year by planting
round it artichokes, sunflowers, or other tall-growing
plants. A way to get rid of the rubbish heap is to dig
a hole in the garden and from time to time put the
refuse in, sprinkling a little fresh slaked lime over
it occasionally. After addition of lime the decaying
vegetable matter is less likely to harbour garden pests.

The raising of currant and gooseberry bushes should
be practised in every school garden, while those who
have the knowledge and opportunities might raise stocks
of fruit trees and practise budding and grafting. The
budding of rose briers is another branch of gardening
that is well worth attention. Cuttings of good varieties
of small bush fruits can be obtained without much
trouble. The best style of bush to adopt for gooseberry, red and black currants respectively, should first
be decided upon. Usually a gooseberry bush has a
stem, twelve to fifteen inches long to keep the branches
off the ground. A red currant bush is of similar form,
while black currants, though often seen on a single
stem, are much better when the branches come up in
the form of suckers. The scholars could be given the
branches of gooseberry from which the cuttings have to

be prepared, and a little questioning should bring out the fact that the buds are branches in embryo and that if a stem has to be made then some of the buds must be removed ; likewise the part that goes into the ground must have the buds removed or else the root portion will throw up suckers. Hence a gooseberry cutting should have all the buds removed from the part that goes in the ground (about three inches); all the buds from the part that will form the stem, while the top should be cut off leaving three or four buds which in the spring will break and form branches. Red currants are treated like gooseberries. Black currants need different treatment, as the part that goes into the ground should have the buds left on, while the part above ground (an inch or two) should have most, if not all of the buds removed. Each year a set of cuttings should be prepared by the scholars. At the end of the first year the scholars should try their skill in pruning the bushes. The aim in gooseberry and red currant culture is to make a bush and therefore the branches are so cut off that the buds left on will form new branches. The direction which the new branch should take should be studied and the pruning done accordingly. After pruning the second year a fairly good bush should have been produced, and when this stage has been reached, a bush might be given to each scholar to take home. From this period onwards a set of bushes could be distributed each year. Black currants do not require pruning, but occasionally the old wood should be cut out.

Cuttings of roses are quite easily raised, in fact some roses, like Dorothy Perkins, may easily be raised from cuttings placed in a jam-bottle of water. Put a little wood charcoal in the water to keep it sweet and place the cuttings in the school window.

The raising of one or more unusual vegetables should be undertaken and these should be cooked or prepared by the girls in the cookery class. Spinach, kohl rabi, celeriac, salsify, chou de Burghley, chicory, artichokes, seakale, endive, butter beans, dandelion and others might well receive some attention in this respect.

A certain part of the school garden should be devoted to flowers. This is often interpreted to mean the growing of a few annuals at one end of each individual plot. This is not a good plan to adopt and little likely to teach love of flowers, care of flowers, arrangement o flowers, all essential ideas connected with the flower garden. To put these latter ideas into practice a strip of ground should be set apart for flowers and this piece of garden would be under the care of all the scholars. Annuals and biennials may well find a place in such a border, but perennials should also be raised from seed and planted therein. See that some tall-growing plants are included, such as anchusa, michaelmas daisy, pyrethrum, lychnis, achillea, as these will help to break the flatness and form groups of colour at different periods. Dwarfer growing flowers could include primulas for the shady places, polyanthuses, doronicum, iberis, potentilla, etc. Bulbs, montbretia, gladioli and many others may be gradually added. Where the garden soil is very heavy attention must be paid to improving its texture. In most cases nothing is so good for this purpose as ballast. This is made by burning alternate layers of the clayey soil and small coal in a fire in the garden. A fire of this type might be continued for several weeks, the ashes and burnt earth being eventually spread over

the ground and dug in.   A fire could be made each year
for two or three years until the desired effect upon the
soil had been attained.

Success in gardening depends upon thorough cultivation of the land more than anything else.   Deep
digging is one of the secrets.   Dig the ground at least
two spits deep, keeping the top soil at the top and the
bottom soil underneath.   Of perennial weeds every
fragment should be picked out.   Continuous hoeing and
weeding during the growing periods, thinning and
transplanting at the right time all require attention.

Animal manure must be supplied liberally at the
right season and dressings of lime added from time to
time.   A watchful eye must always be kept for pests,
ever bearing in mind that " prevention is better than
cure."   The season for planting, transplanting, etc. depends mainly upon the weather and condition of the
soil.   When these are suitable take advantage of them.

Finally, in gardening remember that the present
operations will always depend upon how much had
been anticipated and prepared for.   In the garden, as
in life, it is always the future that has to be looked to.

12.   THE SCHOOL BEAUTIFUL

School buildings and premises, even in country
places, are too often dull, dismal and sombre-looking.
The outside does not denote anything pleasant.   Such
conditions, in the majority of cases, could with ease be
altered.   Where there is no school garden there can
often be found a space which with little trouble could
be converted into a rock garden or fernery.   Either of

these could be made on an asphalt yard without in any way interfering with the asphalt. The only things necessary would be stones and soil. In urban districts where stones are sometimes more difficult to obtain, large clinkers can usually be got from a brickyard or factory. The rougher and more rugged-looking the stones or clinkers the better. Mark out a base and build up irregularly to the height required, putting in soil as the building-up proceeds. The ideal rockery would of course be planned to represent natural conditions, e.g. the stones would be placed in the form of ledges to represent an outcrop of rock or something of that nature, such as can often be seen in a railway cutting or a hillside. But this is simply a matter of arrangement. If a fernery only is desired, then the spot chosen should be a shady one while for a rock garden almost any position will do. Even if a rockery is fully exposed to the sun, shady places behind the stones can easily be arranged and thus sun and shade loving plants can be cultivated. The rockery may be clothed with suitable flowering plants by asking the scholars in the first instance to bring specimens of such plants as saxifrages, sedums, veronica, arabis, aubretia, sea-thrift, money-wort or creeping Jenny (*Lysimachia nummularia*). Many kinds of plants can be raised from seed which can be bought in penny packets. When well clothed a rockery in early summer is a glorious spectacle while at all other times it is of much interest. The characteristics of flowers that live in high altitudes (they are mostly dwarf plants which quickly throw up blossom stems with flowers bright coloured the more readily to attract attention, closing quickly when a heavy cloud passes over, etc.); the common names of the plants such as stonebreakers for saxifrages, stonecrop for sedum, etc.—all these are points of more than passing interest.

Once a teacher has grasped the idea of beautifying the school and its surroundings other means of carrying it into effect will occur. A few climbing roses, raised from cuttings by the scholars themselves, rambling over rustic framework or school walls would be highly attractive. Not only will the work of planting such things as these be of interest to the scholars but the impression upon their young minds must be great.

A rustic plant stand or two in which such simple things as nasturtiums could be grown would often be effective if placed in the front of the school. Window boxes too might be utilised in many instances.

Not only should the outside of the school premises be brightened but the inside of the school itself should receive attention. A hanging basket filled with flowers could be hung at the entrance door or porch. The basket could be made by the scholars, lined with moss, and then filled with soil, into which suitable flowers could be placed. Trailing plants of many kinds can be easily raised from seed and among these may be mentioned *Saxifraga sarmentosa*, which is also called "Aaron's beard," or "Mother of Thousands." *Campanula isophylla* is a simple and effective plant. An asparagus fern could also be used for this purpose. Inside the school other hanging baskets could be provided and there can often be found space where a plant such as an Aspidistra or a castor oil plant might be placed. If the scholars can make the necessary flower stands so much the better.

Cut flowers from the garden should be in constant use for decoration of the schoolroom. The scholars should take charge of the plants, arrangement of cut flowers, etc. Certainly very much more might be attempted in the making of the school and its surroundings pleasing and beautiful.

13. TOOLS AND THE TOOL SHED

THE tool shed should be so fitted up that the tools can be readily and easily distributed to the scholars and can also be inspected with ease by the teacher. To accomplish this some system of storing the tools must be brought into use. It is a good plan to have each of the tools stamped with a number. There are, perhaps, ten spades and these would be stamped 1—10 respectively. The hoes and other tools would also be stamped 1—10. On plot 1 of the garden only tools numbered 1 would be used and so on. Thus it would be quite easy to find out who had used a certain tool. Tools may be hung round the tool shed or put in racks. See p. 152. Another method is to place the tools in brackets. Thus all the spades would be kept in one place, all the hoes in another, etc. A seed cupboard is very desirable, in which to keep the seed. It should have different compartments so that any particular seed may be instantly found. Where the school garden, as is often the case, is some little distance from the school, a few pegs for hanging coats on would be an advantage. Boxes for storing seed potatoes would also have to be housed in the shed, and for this purpose a wide shelf would be most useful. In any case have a place for everything and insist upon everything being kept in its right place. Boys can in turn act as monitors of the tool shed and thus the responsibility for its proper condition is thrown upon the boys themselves. A fairly large supply of plant labels is desirable so that the names of plants, crops, etc., date of sowing, etc. may be recorded. An indelible garden pencil is required for writing on the plant labels.

With regard to the tools these are not always of such a satisfactory character as could be wished for. Boys in the garden class are of various sizes and degrees of strength and while some of the smallest boys could not wield without fatigue a full-sized tool, yet some of the biggest boys could do this, especially if the tool is of a light make. Now if all the tools supplied are of one size, either the tool is simply a toy in the hand of the biggest scholars or else it is too big for the smallest boys. The difficulty can be overcome by supplying tools in two or three different sizes or weights, say, for a class of 14, three full-sized spades, three small ones and the rest intermediate in size. This question of size of course refers more especially to the forks and spades. Many school sets of tools do not include a draw hoe, with the idea of encouraging the use of a dutch hoe. Surely this is a mistake, for where the land is heavy a draw hoe can be used to break up the hard lumps, while it is also desirable for earthing potatoes. Some districts have one or more garden tools of a distinctive character that have been found by experience to be particularly suited to the local conditions and it would be a mistake not to recognise such a factor. Teachers should be allowed a certain amount of latitude in the choice of the garden tools used by their scholars.

The care of the tools is an important matter and they should all be kept nice and clean. After use in wet weather the tools might with advantage be rubbed

over with an oily rag. A spade cleaner is useful to
free spades and hoes from soil. These can be made by
the boys. No tools should be put away in a dirty condition. In connection with the tools their adaptability
for certain operations should be noted. Some most
interesting and useful lessons might be occasionally
given dealing with the evolution of the garden tools;
the materials of which they are made ; their shapes ;
and the mechanics of the tools when in use.

14. EXHIBITING

THERE are very few country villages and towns that
do not boast an annual flower show, and in very many
instances the schoolmaster is intimately connected
therewith. Hence special classes for the school garden
boys can usually be arranged. There is, however, no
need to restrict their entries to the special classes, for
if their produce is good enough let them make some
entries in the open classes. Exhibiting at a local
flower show allows of some of the work of the school
being brought under the notice of parents and parish-
ioners.

The growth of vegetables for exhibition is really a
specialised form of gardening, if the best results are desired. The soil has to be deeply dug, properly manured,
and the cropping so arranged that the plants are far
enough apart to ensure each getting plenty of light and
air, and room for growth. All through their growing
period crops for exhibition must have special attention
paid to them. Constant stirring of the soil with the
dutch hoe, and hand weeding between the plants, must

be practised. Certain crops may be given very weak
manure water at intervals and perhaps an occasional
sprinkling of an artificial manure.

To make sure of having such crops as beans and
peas ready at the proper time, planting must be done
in succession. For example, a row of peas might be
sown one week, another row a week later and a third
row one week later still. Another question that must
be considered is the best varieties of vegetables for
show purposes. If a collection of vegetables is to be
shown, then those kinds that would individually gain
the most points should be cultivated. Cauliflowers, if
of good quality, would gain more marks than cabbage.
Peas and runner beans count well in a collection. For
August shows celery should not be grown, as this
vegetable is not fit for use at this season of the year
and therefore its cultivation as a show vegetable pure
and simple should be discouraged as strongly as possible.
The best method of preparing and staging the vegetables
can be determined by studying prize collections in the
open classes at any good horticultural show. Vegetables
for show should be prepared and set up by the boys
themselves.

The work of growing, preparing and staging vegetables for show will teach the boys very many points
concerning what are considered the good qualities of
certain vegetables. It should lead them to see that
quality is the chief desideratum. At the end of the
season comparison may be made of the crops grown
specially for exhibition and those grown in the ordinary
way. It will probably be found that it answers best
to take pains and trouble and thus raise the finest
crops.

An industrial section to the flower show affords an opportunity of displaying other work of the scholars. Teachers might suggest to the Society what classes would be the most suitable. Classes are often included offering prizes for specimens of handwriting, dried wild flowers, map drawing. While these yield a certain amount of interest, yet there are others educationally beneficial to the scholars and more attractive to visitors. Instead of a class for dried wild flowers substitute a class offering prizes "For the best drawings, coloured or otherwise, of stages of growth, of three plants from the garden," or "Illustrated descriptions of any half-dozen wild flowers." Another suggestion might be " The best drawings from nature of wild flowers."

Other classes of a similar nature could also be included, such as :

(*a*) A set of drawings from nature with brief description showing the life stages of any three specimens of animal life found in the garden.

(*b*) Drawings from nature of insect pests found in the garden.

(*c*) Drawings from nature of "friends" of the gardener.

(*d*) For the best garden tool or appliance made in the manual room.

(*e*) For a rustic flower stand or hanging basket.

(*f*) Collection of specimens of small animal life found in the garden.

(*g*) Best bottle of gooseberries bottled by a girl in the cookery class.

All the work involved could be specified to be done

in school. This could easily be arranged in lieu of corresponding work in drawing, etc.

So far exhibiting at the local show only has been considered, but frequently Nature Study Exhibitions are held in connection with the large agricultural and other shows. Such exhibitions, if they can be freely visited by teachers, are of great value, especially to those teachers from country districts, who often have to work without a chance of seeing what other schools are doing. A nature study exhibition should afford many suggestions to teachers as to different ways in which the work may be taken. It is in its suggestiveness that the exhibition is of chief value to the teacher. Where the area controlled by an education authority is large something might be done in the way of arranging for an exhibition of work to be held in different parts of the area at which a conference of the teachers of the district might also be held.

For nature study exhibitions the indoor school work in connection with the garden is particularly suitable.

A word of caution concerning exhibiting and exhibition work may not be out of place. While the work displayed should reach as high a standard of excellence as possible, yet it should be the ordinary school work ; certainly the best examples. If the ordinary school work is not shown a false standard will be set up and more harm than good done. A teacher will readily recognise whether an exhibit represents honest work such as he might get his own scholars to do. If on the other hand he feels doubtful how and under what conditions the exhibit was made, he is likely to refrain from making any attempt at improving his own work in this direction and thus, instead of exerting a stimulating influence, it will have a contrary effect. It cannot therefore be said too emphatically that all specimens of work at Nature Study and other exhibitions should be produced in the ordinary lessons, entirely by the children themselves and under normal conditions. Thus only will the exhibits prove of real value.

In connection with nature study exhibitions at which a large number of teachers are likely to be present, an opportunity is afforded of holding a Conference, when discussion on nature study and allied work might take place and thus afford mutual help. It is a matter of surprise what difficulties and doubts can be removed by means of a little free interchange of thought. At some Conferences I have attended offers were made by prominent naturalists and others to lend teachers, who cared to apply to them, works on natural history, etc. Offers were also made of identifying specimens, etc. Nature Study Exhibitions should not be competitive as the conditions under which the exhibits are produced must be extremely varied in character and nothing good comes of setting up school against school. There is of course no harm in offering certificates of merit to all exhibits considered worthy. Successful exhibitors should be content to carry off their honours in a quiet manner and without undue advertising.

15. EVENING CONTINUATION SCHOOLS

GARDENING is usually a popular subject with evening students. Theoretical lessons during the winter are supplemented by practical gardening in the spring and summer. Gardening may be taught to evening scholars as a technical subject, and therefore there is the greater need for the work to proceed on good lines. In this respect nothing would be better than conducting the evening school garden, as far as practicable, along the same lines as has been suggested for the day school garden ; giving perhaps less drawing and composition. The planning to scale of the individual plots should be the scholar's own production and each scholar should learn to plan and conduct his garden on his own initiative.

Special stress might be laid upon intensive cultivation, without, however, using expensive aids; although the management of hand lights and a frame would be most useful. In some cases these could be made by the scholars themselves and thus would have a greater value.

Particular attention should be devoted to getting as many as possible of the most profitable crops off the ground in the shortest time. Questions dealing with the germinating qualities of seeds might be dealt with, the scholars finding out the respective values of different samples. The germinating value of seeds may be found as follows :

From a sample of seed count two lots each of 100 seeds. Place each lot between a folded piece of damp flannel or blotting paper and then place on a flower-pot saucer stood upside down in a shallow vessel of water to ensure the seed being kept moist. The whole should then be covered by a seed pan or large flower-pot and kept in a fairly warm room. Each day the seeds which have germinated may be taken out and record kept of the numbers. If the seed is of good quality the germinating should be fairly even and regular. When the test has been completed the average of the two lots can be found and the germinating capacity per cent. This experiment has not taken any account of the purity of the seed, which of course is a factor in finding the real value of the seeds. Where the purity of the seed can also be found as well as the germinating capacity, then

$$\text{True value of seed} = \frac{\text{Germinating Capacity} \times \text{Purity}}{100}.$$

For instance, suppose germinating capacity is found to be 84 % and purity of seed 90 %, then

$$\text{True value} = \frac{84 \times 90}{100} = 75\cdot6.$$

Thus 100 seeds would only produce 75·6 of good seed.

The methods of simple forcing of rhubarb, growing of seakale, asparagus, etc. might receive some attention. If many of the scholars are likely to be engaged in gardening as an occupation, then the packing and grading of fruit and vegetables and the materials (tissue paper, wood wool, etc.) used for packing could form a subject of study[1] and this would naturally lead to the question of markets, the sizes and quantities of goods usually acceptable. Further work might deal with the question of the continuance of supply, of branding to show quality, of the value of co-operation in the collection of small lots of produce for sale and also in the purchase in bulk of seeds, fruit trees, artificial manures and other garden supplies. The methods of keeping accounts; of remitting sums of money and the means taken to ensure

---

[1] *Packing of Apples in Non-returnable Wooden Boxes.* Leaflet by South Eastern Agricultural College, Wye, Kent. 6d.

the safety of same in transit; trade and other references
and the investigation of same; different classes by which
goods can be sent by rail and the charges for same;
a banking account, how it is opened and its uses and
advantages; the wording of telegrams; telephones and
their uses; the writing of advertisements offering produce for sale and the most suitable periodical in which
to display the same; the answering of advertisements;
the computing of discounts; working out profit and loss
accounts; invoices, bills, receipts; rates and advantages
of insurance of buildings, appliances, horses, etc.; life
insurance and its value as a security—these are among
the many subjects that may be closely connected with
evening class work in gardening. The study of these
matters should prove highly attractive and educational;
much more so than the syllabuses of elementary mathe-
matics now enforced in some districts for evening
scholars. Evening school work to be of any real value
must deal with life. Further work that might be useful
as well as educational could be given dealing with
emigration, where to find out reliable information of
any country, routes, cost, conditions of life in the new
countries, the kind of employment likely to be in
demand, and more or less the geography of countries
to which emigrants usually flock. The packing of flowers,
plants, bush and other trees for transit, also the study of
artificial manures, washes and insecticides might form
other useful practical work.

Woodwork could be undertaken and the scholars
might, without a very extensive or expensive equipment,
make a number of horticultural and other appliances.

It may readily be seen from the suggestions given
that a useful course of study extending over two or

more years might be selected. Not only would gardening operations be taught, but knowledge would be imparted of a most useful and varied kind, especially valuable to those who would spend their lives mostly on the land, either in this country or in one of our colonies. The training throughout would be given with the view to making the youths self-reliant, to showing them how and where to obtain knowledge as well as how to use it.

16. WOODWORK AND METALWORK

MANY schools taking gardening have manual training classes in woodwork which their scholars attend. If that is the case then there are a large number of objects, tools, etc., which the boys can make in lieu of the ordinary scheme models. Objects made by the boys and afterwards used by them have a value greater than models, which after all only remain models. To such schools the drawings and descriptions of tools, apparatus, and appliances given will prove suggestive. But it is not entirely for the schools that have every opportunity of working in wood that this chapter is written. Woodwork proper is only introduced into those schools where a grant can be earned and where a teacher has special qualifications for teaching the subject; but there are many schools where the teacher has sufficient knowledge of general woodwork and where with a small equipment some very interesting and useful work could be undertaken in this direction. The outfit must of course depend somewhat upon what it is proposed to attempt and the number of scholars catered for at any one time. A bench fitted with a vice is a necessity, but this need not

7—2

be an expensive affair—a bench 18 to 20 inches wide and about 4 to 5 feet long would prove a suitable size. In some cases a strong box, an old table or discarded school desk might be utilised for a bench. Where room for working is very restricted it might be an advantage to have a collapsible bench which could be stored away when not in use. A saw stool or an equivalent is another necessary article. A hand saw (22 inches long is a suitable size), a jack plane, hammer, steel rule, pincers, knife, bradawls, brace and bits, gimlets, a try square, and an axe of American pattern might be looked upon as a minimum equipment with which to make a start, and other tools could be gradually added as required. These would comprise two or three chisels, a bow saw, tenon saw, wood and metal files, marking gauge, screwdriver, and mallet. Thus the cost of equipment would not be great. The wood for use could be purchased in the form of empty packing cases, etc. In some districts much suitable wood can be had at the price of firewood that would do very nicely for making rustic and other objects. Trimmings from larch poles, etc., are particularly suited for this purpose, while prunings of apple trees, hedgerows (especially hedge maple), thin ends of silver birch can also be often obtained.

The most difficult questions however will not be the provision of tools and appliances but the conditions under which the work shall proceed. Many schools are now experimenting in the direction of devoting three afternoons a week to practical work, and where this is the case woodwork can be taken in the schoolroom. In other cases there are difficulties to be overcome. A classroom is usually out of the question, if the work proceeds during school hours as it should do, owing to the noise disturbing the other scholars. The question of superintendence of the work will in many cases cause the most trouble, because with a small equipment it is not possible for all the scholars to be at work at one time. If light woodwork is taken in the school, then a portion of the class could be doing this and another part doing heavier woodwork. Many of the articles described may be made as "light woodwork" objects, while a few could be made with a knife only. Good workmanship should be insisted upon as far as possible, as the woodwork would be introduced to give training to hand and eye, and this means making things as accurately as possible. Every article should be executed from either a rough sketch or from a scale drawing made from a similar object or from data supplied. If the object is made from a rough sketch, then a proper scale drawing should be prepared when the object is completed. By connecting the drawing in this manner with the woodwork the work should become more careful and exact. Home-made tools and appliances help to bring out and develop the inventive faculties, and assist the scholars by making them rely upon themselves. Tools, like garden reels, dibbers, etc., might be presented to the scholars when they leave school provided they have acquitted themselves satisfactorily in their school career. This would necessitate new sets of these tools being made each year, and thus each new set of boys would have an opportunity of making some of the tools. In some cases the objects may be improved by the use of metal or other material and if so these should be made full use of. Articles should be finished by coating with paint or other preservative such as creasote. Where metalwork can be taken many of the garden tools, such as trowel, reels, hoe, Dutch hoe, etc., may be constructed.

Objects for use in the garden will naturally be exposed to the weather and hence a suitable wood such as yellow deal should be used. Bass-wood, yellow pine, and satin walnut are not suitable woods for making plant markers, pea guards, etc.

Several of the objects described are not suitable for individual work in a manual room because they would take a boy too long to complete properly. Where this is the case and it is desired to make such an article it can be performed by co-operation. The class may first of all discuss the article to be made, and each member may prepare sketches and particulars. After an examination of these and a further discussion of points raised by the drawings, a boy who has given the best and most complete particulars should be given charge of the making of the object, for which purpose he would select a number of boys to each of whom he would allot a portion of the work. Each boy would thus have to make his work fit in with that of the others and it would be seen how faulty work in one instance would spoil all the rest. Co-operative work thus conducted has much to recommend it.

Plant Markers

A plant label is a well-known model suitable alike for light woodwork and the manual class. Instead of giving a set shape for this model, let the class see a plant label and then discuss its use. They will thus easily discover the essentials of the object. (1) A part more or less pointed to stick in the ground easily. (2) It must be of sufficient size to stand out above the ground so as to be seen readily. (3) It must have a part where the name can be written. (4) As the label will have to stand out in all weathers the top should be so made as to shed the water as much as possible.

Having found out these essentials the class might suggest different ways of shaping the pointed end and of the top. This done, let each boy make a rough sketch

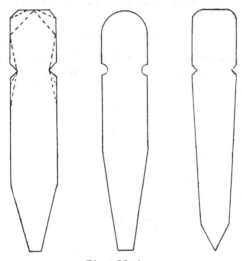

Plant Markers

of the shape he would like to make and the dimensions he would use. Various shapes are illustrated.

This method of dealing with the models should be largely employed, and rightly used must have a very stimulating effect upon the scholars.

Plant labels for use in flower-pots, for tying to fruit trees and a variety of other purposes, may be cut out of sheet zinc, a pair of tinman's snips and a file supplying the necessary tools. These labels may be made in many different shapes and should be designed by the scholars. Most seed catalogues give illustrations of the types of label now described. Names are written on them with a solution of copper sulphate and a quill pen. They form more permanent labels than the wooden ones.

Layering Pins may likewise be made of zinc. They are used for fastening down carnations and other plants.

A Plant Support

A plant support is a suitable exercise for light wood-work. It is better to screw the strips together than

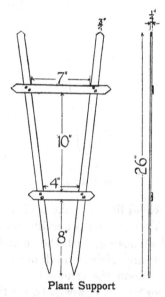

Plant Support

nail them. The sizes and shapes must depend upon requirements.

Garden Plot Marker

In school gardens it is advisable to mark off each plot at the corners by means of stakes. A plot stake is shown, but the top instead of being chamfered might be

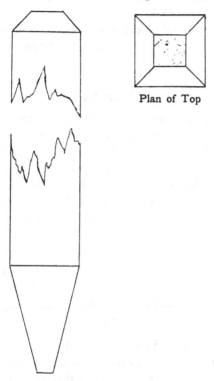

Plan of Top

Garden Plot Marker

rounded. The stake itself might be rounded instead of being made square. Before use it should be painted or given a coating of a preservative such as creasote. A suitable size is about $14'' \times 2'' \times 2''$.

FLOWER STICK OR GARDEN STAKE

Flower sticks will of course vary in size according to the purpose for which they are to be used. A small one for flowers may be half an inch to three-quarters of an inch in thickness, while a suitable one for holding a rose tree in position would have to be about $1\frac{1}{4}$ inches square. Before being used coat with green paint or a preservative.

A WEBBER

A webber for putting cotton over fruit bushes is shown. It consists of a square or rounded piece of wood about three feet long by an inch thick. A slanting hole is drilled in one end with a $\frac{5}{32}$ inch drill bit. About six inches from the other end a nut and bolt with two washers are fixed. About half-way between this and the top end a staple or screw hook is fixed. A reel of

A Webber

black cotton is placed on the bolt with a washer on each end of it and the nut screwed on ; a wing nut is best for the purpose. The cotton is threaded through the staple and the hole at the top. In using the webber the end of the cotton is first made fast to a twig and then the cotton can with ease be threaded where desired by simply moving the webber.

This model could be made as "light woodwork."

Spade Cleaner

The spade cleaner is an interesting and useful object and one that is quite simple in form. A man engaged in digging frequently wears one of these just below the knee, so that it is readily available for cleaning the spade. It is an object that would be useful for cleaning

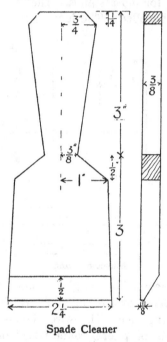

Spade Cleaner

the garden tools before storing them away. If need be the handle can be made longer and a hole put through it so that it can be hung in the tool shed. The shape may be varied by the shoulders being rounded. Scholars should make their own designs.

Stand for Museum and other Purposes

The stand consists of a wooden base and wire supports. The wood should be about half an inch thick, yellow deal, bass-wood, or yellow pine being suitable. The length may be varied as necessary. It may be made to accommodate one or more tubes. A width of $2\frac{1}{2}''$ is sufficient for most stands. The front edge may be rounded or bevelled. The loop in the wire is made

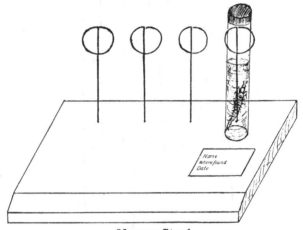

Museum Stand

by twisting a piece of wire tightly round the handle of a bradawl, or chisel, or a round stick, which needs to be firmly held in the vice. The wire should be bent by means of a pair of pliers, round-nosed ones for preference. Having made the loop this is bent at right angles to the stem. Wire that is fairly stiff should be used. If preferred the wooden stand may be covered with white paper on which may be written the name of specimen displayed.

A Plant Label

This plant label demands more skill in making than does the ordinary type. The name part may have the corners rounded or cut to different angles than those

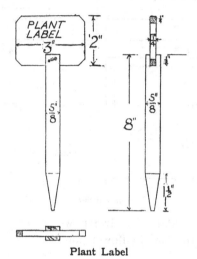

Plant Label

shown. When finished the stem should be painted with a preservative, and the name part given a coating of white paint.

A Soil Firmer

When sowing seeds in boxes and pans it is often necessary to firm the soil somewhat, and for this purpose a soil firmer is a useful appliance. In addition to being utilitarian it is an excellent model for the manual room. The shape and size of the base piece may be varied—square, oblong, elliptical, or round.   The handle, too, lends itself to being varied considerably.

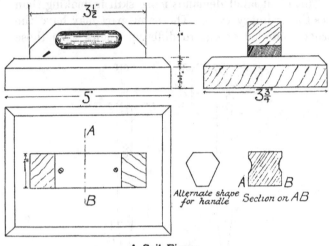

Alternate shape for handle   Section on AB

A Soil Firmer

A small circular soil firmer would be useful for making firm the soil in flower-pots.

A Soil Treader

Frequently in gardening operations it is necessary to have the soil nice and firm, e.g. the onion bed.   For this purpose a pair of treaders, to be worn on the feet, may be made.   Each is made from a piece of yellow deal about 15 inches long by 7 inches wide and 1 inch thick.   Across the front a piece of old leather strap is looped and nailed in position.   Into this loop the toe of the wearer fits.   A piece of wood, shaped as shown, will help to keep the heel in place, while a strap may be fixed which will buckle round the ankle.

The treaders are used by first securing them to the feet and then walking up and down the piece of land

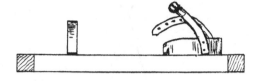

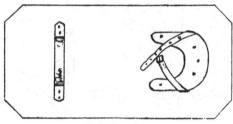

Soil Treader

to be firmed. It is an advantage for the user to carry a stick upon which he can support himself when turning round.

A Hanging Flower Basket

A hanging flower basket would be a suitable object for light woodwork or a manual class. It may be made of strips of deal $9'' \times 1'' \times 1''$. The strips may be threaded on wires. The bottom strips need only be half an inch thick.

A hanging flower basket, however, looks much more effective and artistic if made from small branches of hazel, hedge maple, etc. The sticks for this purpose should not be closely trimmed, so that the basket may

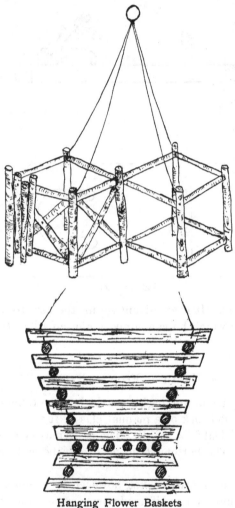

Hanging Flower Baskets

be as rustic-looking as possible. The shape of the
basket may be square, hexagonal, etc.

Another type of hanging basket may be made by
covering a box with virgin cork.

The baskets look nice if moss is laid round the sides
and bottom before soil is placed in them.

Suggestions for hanging baskets are given in the
illustrations.

DRILL MAKERS

In planting seeds a drill maker is most useful.
A simple form is shown in No. 1, while in No. 2 a more
elaborate form is depicted which will draw two drills at
one operation. The teeth may be of various sizes and
designs. Some suggestions for shapes of teeth are
given.

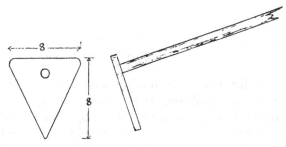

Drill Maker. No. 1.

A drill, too, can be made so that the teeth are adjust-
able to different widths. For this purpose holes would
have to be bored in the stock and the teeth fastened
through these by means of nuts and bolts. If wing nuts

B.   8

are used then the teeth can be moved without the use
of a spanner.

Suggested shapes of Teeth.

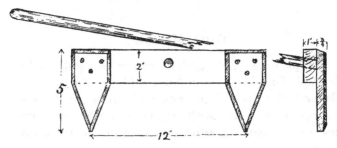

Drill Maker.   No. 2

Seed Box

A box for raising seedlings indoors is almost a
necessity where gardening is practised. Fairly strong
boxes should be made, suitable sizes being $12'' \times 9'' \times 3''$;
$15'' \times 10'' \times 3\frac{1}{2}''$. A handle can be provided similar to
that given for potato box, p. 115, but in this case it will
be better for the handles to be nailed to the outside of
box or else there will be difficulty in covering box with
glass, etc. A few holes should be bored in the bottom
to afford drainage. Old packing cases may be used for
making seed boxes.

Potato or Fruit Box

This potato box measures $18'' \times 12'' \times 3''$ and is made of wood half an inch in thickness. The handle supports are $6'' \times 1\frac{1}{2}'' \times 1''$ and the rails forming the handle are

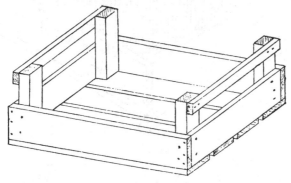

Potato Box

$1'' \times \frac{1}{2}''$. Of course the box could be varied in size or in details. The wood should be half an inch thick. Note that the bottom strips are apart a little.

A Sieve for Potting Soil

The size of the sieve may be varied. Different forms of handles may be adopted, of which two types are illustrated at *A* and *B*. The bottom is covered with wire netting of half-inch or smaller mesh. The wire bottom is held in position by being nailed through strips of wood as shown. This is better than fastening with nails alone or staples.

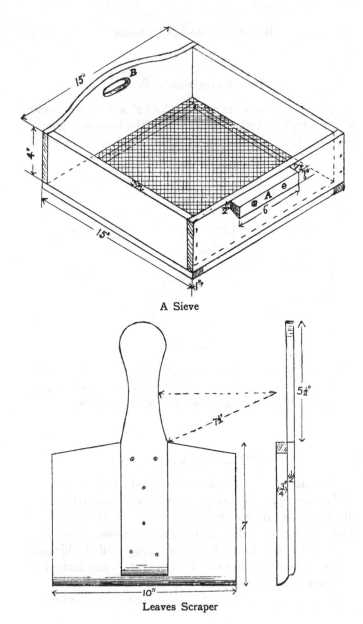

A Sieve

Leaves Scraper

Leaves or Rubbish Scraper

A pair of these is used for gathering up leaves, etc. One part is held against the leaves and the other part brought up to this and the leaves held between them and thus lifted into a barrow, etc. The handle and the shape may be of different designs.

Bird Clappers

Bird clappers form a popular model which is capable of being executed in light woodwork. Three pieces of wood are tied rather loosely together. A piece of old

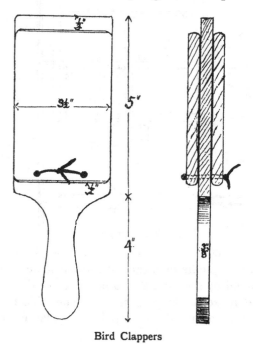

Bird Clappers

leather shoe-lace serves this purpose. The handle may
be designed.

A Wind Clapper

A wind clapper is a rather more elaborate affair
than hand clappers. A simple frame is first made about
10 to 12 inches long and about 8 inches high, and this is
made into two equal divisions as shown. A spindle of
ash of $\frac{3}{8}$-inch diameter goes through $\frac{1}{2}$-inch holes and
carries the sails, the clappers and the balancer. The

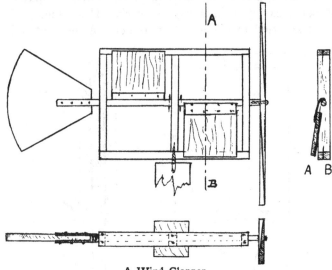

A Wind Clapper

clappers, which come a little below the frame, are
nailed to the spindle by means of leather hinges. When
driven round by the wind the clappers strike the framework and make a clapping noise. The object when
completed is mounted on a pole and placed in the
garden.

A Hand Barrow

A barrow such as the one illustrated is a very handy appliance and in many ways more useful in the school garden than either a wheelbarrow or a truck. A good strong box of suitable size should first be selected, and this should be strengthened either by nailing strips

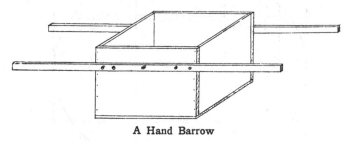

A Hand Barrow

across the bottom, or by means of narrow iron banding. The handles should project about 2′ 6″ to 3′ at each end and should be shaped or rounded somewhat where held for carrying. Ash is the most suitable wood to use for the handles.

A Truck

A truck may be made from an old packing case and used instead of a wheelbarrow in the school garden. The wheels may be made of deal or elm.

Pot Stand for Bottled Fruit

Nail together four strips of wood about a quarter of an inch thick, by means of two cross pieces. Mark out the shape and cut out with a bow saw. The stand should be made to fit loosely a pot or other cooking vessel. If a bow saw is not available a serviceable stand can be made by sawing the corners off.

For bottling fruit the stand is put inside the pot with the flat side upwards. On this stand the bottles of

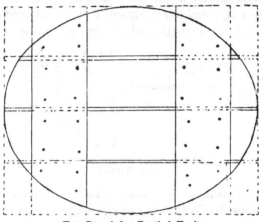

Pot Stand for Bottled Fruit

fruit to be sterilised are placed and then there is no need to use hay, cloths, etc. to wrap round the bottles. The spaces between the strips of wood allow of circulation of water. Wood from a packing case is suitable for making this model.

JAM STIRRER

A jam stirrer is a useful article for home use or for the cookery class. At the same time it forms an excellent woodwork model. It should be made of beech, sycamore or American white wood. It lends itself to being designed and treated in different ways. For instance the blade may be scooped out a little, etc.

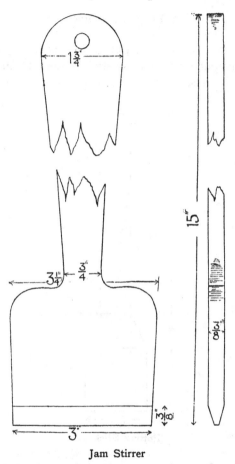

Jam Stirrer

A POTTING STICK

Many plants have to be put in their pots very firmly, and for this purpose a potting stick is most useful. The base may be a square or rounded piece of wood, into

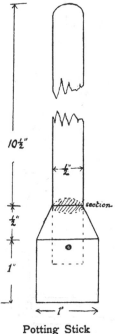

Potting Stick

which a handle is fixed and held firmly by a wooden peg. The potting stick could also be made of one piece of wood.

A Boot Scraper

A boot scraper for use in the garden is almost a necessity, especially in wet weather. No. 1 is made of two uprights of ash, rounded or left square, each about $16'' \times 1\frac{1}{2}'' \times 1\frac{1}{2}''$ or thereabouts. The tops are rounded

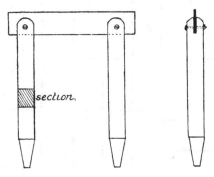

A Boot Scraper. No. 1

to throw off the water. The bottom ends are pointed to facilitate driving into the ground. A piece of sheet iron, or part of an old hoop, forms the scraper. The wood should be coated with a preservative before being fixed in the ground.

No. 2 boot scraper is made of a framework and lengths of stout iron hoop, or better still of mild steel about $\frac{1}{8}$-inch thick. The sides are $10\frac{1}{4}$ inches long by 2 inches wide and $1\frac{1}{4}$ inches thick, while the ends are 8 inches by $1\frac{1}{4}$ inches by $1\frac{1}{4}$ inches. Half-lap joints are used. Yellow deal is a suitable wood. The iron strips are fixed in saw cuts made in the frame to a depth of

about ⅝-inch. The tops of the strips should be about
½-inch above the framework.

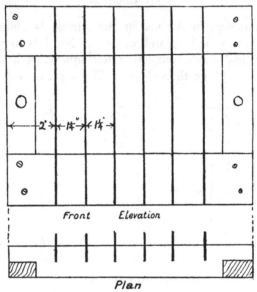

Front   Elevation

Plan

A Boot Scraper.   No. 2

A DIBBER

A garden dibber is a model that may be made in
many forms. A serviceable one can be made from a
forked stick of suitable size and shape cut from a hedgerow. The top part should be forked to fit the hand.
While most dibbers are round in shape, there is no
reason why the stem should not be square or octagonal.
A dibber could be made without a handle but would be
less effective in use than one properly handled. Several
forms are illustrated, including one specially designed

for planting potatoes, leeks, etc. This latter is driven
into the ground by the foot. It will be noted that
different methods of joining handle and stem together
may be employed.

No. 1. This dibber can be made by boys who are
not able to make halving or mortise and tenon joints.
It can be shaped with a pocket knife. The hole should

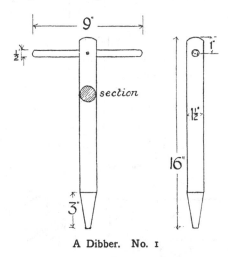

A Dibber. No. 1

be bored in the stem before the wood is rounded.
A smaller dibber of the same type could be made for
pricking out small seedlings.

No. 2. A dibber of simple form. The handle is
bevelled but might as easily be rounded. The stem
could be square or octagonal. Half-lap joint is used.
The sharpened end might be rounded.

No. 3. This shows a dibber with holes bored in the
stem through which a stout nail or wooden pin can be
put and thus the depth to which the dibber is driven

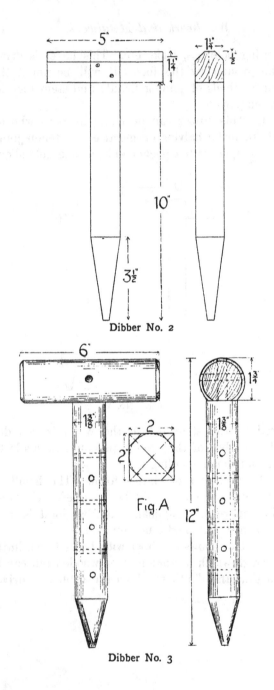

5"

1¼"

1¼"

½"

10"

3½"

Dibber No. 2

6"

1¾"

1⅜"

2

2

Fig. A

1⅜"

12"

Dibber No. 3

into the ground may be regulated.   Fig. A suggests how
the handle may be shaped.   Holes of ½-inch diameter
are bored at inch distances apart.

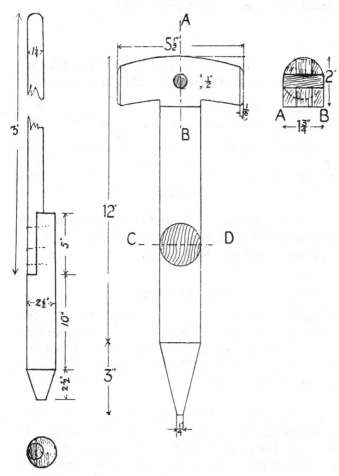

Foot-driven Dibber No. 4   Dibber No. 5

No. 4 is a dibber driven in with the foot while No. 5 is an ordinary type.

A dibber may be made more effective by having its pointed end covered with a piece of zinc.

A Rake

A front and a side elevation of a garden rake are shown. The stock may be rounded or kept square and

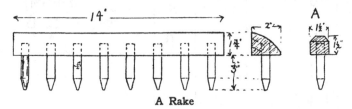

A Rake

bevelled as shown at *A*. Instead of wooden pegs, stout wire nails, French pattern, may be used. The stock and teeth may be varied considerably.

Garden Reels

Garden reels can be made by simply pointing two sticks with a pocket knife, but more elaborate ones form very popular models in the manual class. Two forms are illustrated, of which the first can be made as an exercise in light woodwork as well as in the manual class.

No. 1. This consists of four strips of flat wood, screwed or nailed together as shown. To complete the model another strip of wood must be prepared, similar to the back piece, except that it has a hole bored in it near the top for the purpose of tying the string thereto.

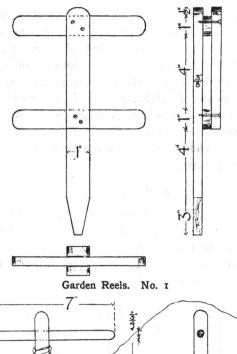

Garden Reels. No. 1

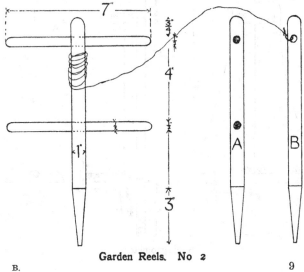

Garden Reels. No 2

No. 2. This is made of rounded wood. The holes should be bored before the rounding of the wood is commenced. Ash or yellow deal is the most suitable wood to use. Of course the pins might be made square or octagonal in shape. The cross bars, too, may also be fitted by means of halving joints. Two views are shown of *A* and one of *B*, which form the respective ends to which the line is affixed.

The lines should be given a coating of a preservative before being used.

A Hygrometer

A hygrometer upon the principle of the cuckoo clock can easily be made. The design for the back and shelves may be altered as desired. From the top shelf a piece of catgut is hung, the bottom of which is fixed

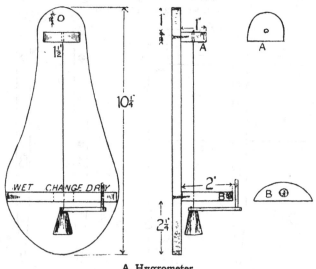

A Hygrometer

in a piece of lead. A pointer is nailed to the lead and this can then revolve round the circular shelf *B*. The catgut is fixed tightly in the top shelf by means of a piece of match stem. Views, but not to scale, are given of shelves *A* and *B*. The shelves are screwed to the back. The hygrometer when completed is hung out of doors and the positions of the pointer for wet and dry weather respectively noted for some time and afterwards marked on the front of shelf *B*. It will be found that the pointer keeps to one side of the shelf for wet weather and to the other side for dry weather.

WEATHER GLASSES

A weather glass which will indicate weather changes can be made by dissolving in 2 ozs. each of pure alcohol and distilled water the following ingredients : 2 drams camphor, and half a dram each of ammonium chloride and potassium nitrate. When thoroughly dissolved pour the solution into a large test tube, eight inches long and one inch diameter. Cork the tube, which should then be either mounted on a board or else fixed in a stand. The making of a back with brass straps or a stand will call forth the designing powers of the boy. Certain modifications of the liquid will indicate weather changes, these should be found out by the boy and thoroughly tested. They might then be neatly written or printed and hung near the instrument.

Another type, which is of an even simpler nature than the above, may be made of a piece of small bore glass tubing, about 3 feet long, and a bottle fitted with a sound cork. About one-third of the bottle is filled with water coloured with red ink. The glass tubing must be inserted through the cork until it dips a little below the surface of the liquid. Cover the cork with paraffin wax or with sealing wax dissolved in methylated spirit to render it airtight. The apparatus is then placed inside a larger jar and packed round with sawdust. Blow down the glass tube until the coloured liquid ascends the tube for about 8 inches. A strand of thin brass wire may be fixed round the tube to mark this position of the liquid. The connection between the height of the coloured water and the weather should be determined by observation. The apparatus may be fitted to a stand, designed and made by the scholar.

A WIND VANE

A wind vane is a familiar object in many country gardens. It assumes a number of forms but all work on

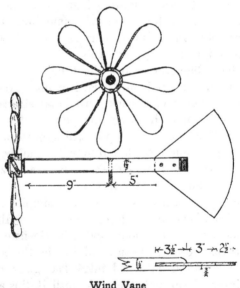

Wind Vane

the same principle. It is a first-rate model for a boy to make because it will "go." Boys should devise their own model. The screw shown in the illustration is where the vane is screwed to a post. A washer is put between the head of the screws and the sails and also between the sails and the shaft.

BUTTERFLY OR SPECIMEN CASE

A specimen case for butterflies and other collections is not beyond the powers of a scholar attending a woodwork class. It usually proves an attractive model. The

Butterfly Case

size may be varied. It will look nice if the sides and top are made of mahogany. The bottom may be of three ply wood, which an old tea chest will supply. The sides may be nailed together, trenched as in the illustration, or dovetailed. Instead of making a hinged frame for the top, the sides are rabbeted to carry a sheet of glass which is held in position by means of half rounded strips of mahogany mitred at the corners and screwed on after the glass has been put in place. Instead of a rebate being made, if too difficult, a thin strip of wood can be nailed on the inside of the box to carry the glass. To open the box the strips would have to be unscrewed. A plan is shown with the glass and top strips removed and a sectional elevation is given showing glass and frame in position. Size illustrated is $12'' \times 10'' \times 1\frac{1}{2}''$. The frame is made of strips $\frac{3}{4}'' \times \frac{3}{8}''$.

DISPLAY CASES

A strip of yellow pine, $16'' \times 1'' \times \frac{1}{2}''$ trued up to $\frac{3}{4}''$ wide and $\frac{3}{8}''$ thick, should make a case about $5''$ by $4''$. A marking gauge may be used as a cutting gauge with which to make a rebate on each of the two inner edges. This should be cut about an eighth of an inch wide by a twelfth of an inch deep. Finish the rebate with a bevel-edged chisel, using it as a scraper. Mark out the four sides, noting that two pieces the thickness of the rebate have to be removed from each of the two long pieces. See Fig. 2.

When the pieces are ready they may be nailed together. The inside of the wood should be coloured or stained black. Two pieces of 15 oz. glass may be cut to size. The cases might be made to take old half or

quarter plate negative glasses. Make one of the glasses
quite clean and then mount the objects to be displayed
thereon, using freshly made glue into which a little
preservative such as carbolic acid or formalin has been
put. Arrange and mount carefully, using the smallest
possible quantity of the adhesive. Glue into one of
the inside corners of the case a small piece of naphthalene. Set the glass aside in a dry place for a day and if

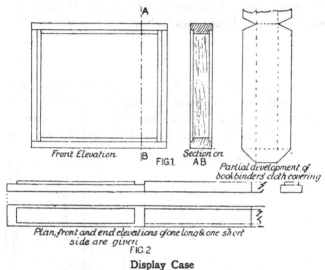

Front Elevation |B   FIG.1.   Section on AB

Partial development of
bookbinders' cloth covering

Plan, front and end elevations of one long & one short
side are given
FIG.2

Display Case

the mounting has been successfully accomplished the
case may be completed. A piece of bookbinder's cloth
is used to cover the framework and hold the glasses in
place, using the same adhesive as before. Cut the
cloth about half an inch longer than the four sides and
wide enough to allow a quarter of an inch overlap top
and bottom. At each of the eight corners the cloth
will have to be cut so as to allow it to fit and is then

well covered with thin glue or paste and carefully stuck on, thus completing a most attractive-looking display case. A narrow strip of white paper may be gummed along one edge and the name of specimen and other particulars written on. Another method of labelling is to put the name strip inside the case. Specimens thus mounted, showing complete life stages, may be handled without much fear of damage. They are kept free from dust and mites. Both front and back of the creatures may be seen and the cases look well in the museum.

Larvæ Box

Where the breeding of grubs is undertaken a larvæ box of some kind is almost a necessity. A simple type is illustrated which is made from a box and a sheet of glass that can be drawn backwards and forwards to

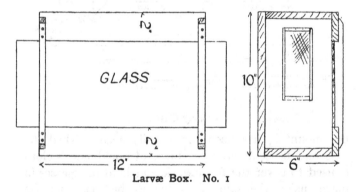

Larvæ Box. No. 1

make an opening into the box. On one of the sides a hole is cut and covered with perforated zinc so as to give ventilation.

The whole front may be made of glass by nailing on

strips and thus forming grooves on each side in which the glass would run.

A larvæ box, No. 2, of a more complete character may be constructed in accordance with the drawing shown. The bottom part of the box consists of a floor raised about 2½ inches high. The front of the box has two glass panels, while the ends are covered

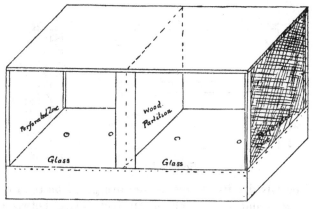

Larvæ Box. No. 2

with perforated zinc. A partition of wood divides the box into two portions. A door to each compartment is made in the back. In the floor one or two small holes are made through which stems of plants on which the grubs feed may be put into vessels of water and so kept fresh. Plug the holes with cotton wool so as not to let the occupants escape. A suitable size for the box is 20 inches long, 7 inches wide, and 10½ inches in height.

SETTING BOARDS

Front and end elevations of a board for preparing butterflies, etc. for mounting are shown. The dimensions are 14″ long by 2½″ wide. The base piece is of yellow pine or American white wood planed to a quarter of an inch thick. Upon this is glued a piece of cork lino. This should be put under a weight and when

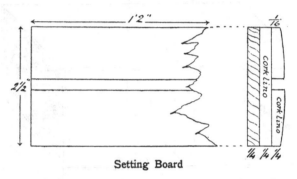

Setting Board

thoroughly dry two strips of lino are glued on the top, leaving about a quarter of an inch space between. (Smaller boards would have smaller spaces.) Place under a weight and let stand for 24 hours. By means of file and glass paper the top surface is slightly rounded. If desired the board may be covered with a thin white paper.

OBSERVATION BOX

In studying some forms of small animal life it is an advantage to be able to observe fully the movements of the creature when it is in surroundings that are as natural as can be made. The difficulty, of course, is due to the opaqueness of the soil and this can only be overcome by putting the creature in a thin section ; hence the design of this box. It will be seen that there are two compartments separated by a porous tile. The narrow compartment has a glass front and in it is placed soil, in the top of which some grass or other plants may be grown. The creatures to be observed are then placed in the soil and their movements and

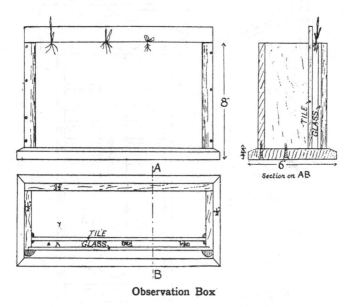

Observation Box

actions can be readily seen. A sheet of brown paper should be put over the glass when observations are not being made. To keep the soil from becoming too dry, damp moss may be kept in the hind compartment and the moisture will then find its way through the tile. The glass may be fixed in position in the same manner as shown for the butterfly case, p. 133.

A Thermometer Stand

A thermometer stand to hang in the playground or school garden is shown in front and side elevations. It is made so as to allow a free circulation of air around

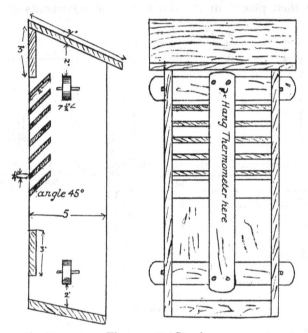

Thermometer Stand

the instrument. If necessary a door could be made of laths like those of the back and then the thermometer could be kept under lock and key. Other designs may be made as alternatives to this model.

Plant Protector or Frame

The plant protector is made of two triangular shaped end pieces 1′ 6″ wide at the base and about 1′ 6″ high. Matchboarding $\frac{3}{4}$″ thick is suitable to use. The ends are fastened together by a triangular strip which projects about 8″ beyond each end. These projecting parts may be shaped and thus form handles for the carrying of the frame. The framework carries two lights, only

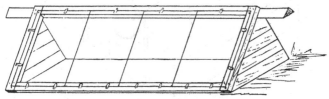

Plant Protector

one of which is shown in the drawing. The frame for the light is made of wood 2″ × $\frac{7}{8}$″ and the ends are half lapped. A rebate may be cut for the glass which may then be puttied in, or else the glass can be held in position by means of wooden buttons or metal clips. Length of frame 4 to 5 feet, but it can be made to any suitable dimensions.

Moveable Frame or Hand Light

This frame is made of $\frac{3}{4}$-inch wood, but the ends may be made of stouter material. The handles of thin iron rod may be shaped by the boys and fastened by means of staples. In the end elevation given the position of the glass is illustrated. One piece projects over the ridge, and the other piece does not reach to the ridge. Thus ventilation is afforded. The clips to hold the glass in position can be made of lead or sheet zinc, and should be bent as shown. This is a most useful form of frame.

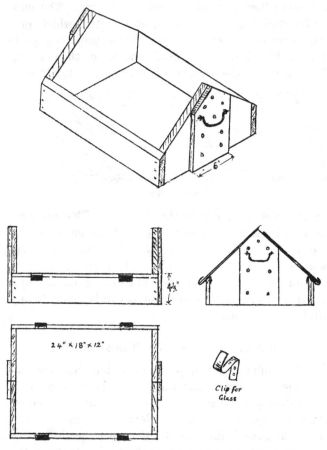

Plan, Front and End Elevations and Isometric Projection
Moveable Frame.   No. 1

Another similar form of frame is shown in No. 2. The details of this are shown in the drawing. The zinc

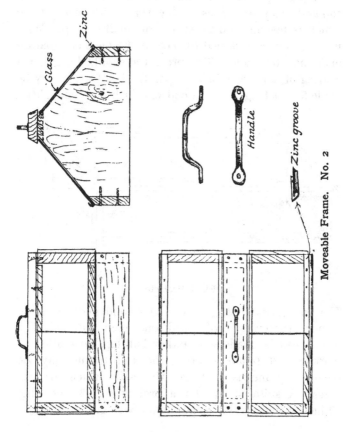

for the bottom of the glass is carried the whole length of the front. This should be given a slight fall towards one end to allow of water being thrown off.

A Plant Protector

A plant protector is illustrated which is made of a top and two side frames, each covered with wire netting which is fastened on by means of small staples. Wire netting can be fastened to each of the ends or separate ends may be made. The protector should be given a coating of a preservative or paint. The frames may be made 3 feet long and of wood one inch by five-eighths.

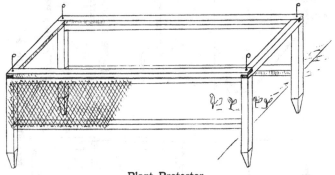

Plant Protector

The end posts that go into the ground may be made rather stouter, say of wood one inch square. The frames can be made with either half-lap, dovetailed or mortise and tenon joints. The top frame may be secured by means of large wire nails which fit into holes. A suitable size for a protector is 3 feet long, 12 inches wide, and 8 inches high above ground level.

A Fruit Drying Tray

In districts where fruit is plentiful girls might in their cookery lessons be taught to dry fruit by evaporation. The process of fruit drying is not a difficult one.

Apples and plums are the best fruits for drying. Apples
to be dried should be cut into quarters and cored and
then placed on drying trays and put out into the sun for
several successive days. At night the tray should be
put into a warm oven with the door not quite shut

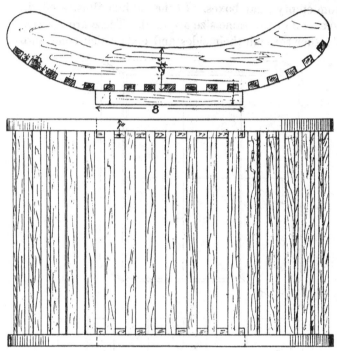

Fruit Drying Tray

to allow moisture to pass out. The drying may be done
entirely in the oven but very little heat must be used
and the door kept open. The tray for drying fruit may
vary in size and shape. Instead of the bars being let
into the side pieces they may be nailed or screwed on.

The tray is designed so as to allow of a free passage of air. It will of course be made of a size to fit the oven.

A Seed Cabinet

A simple but effective seed cabinet may be made from empty cigar boxes. In the cabinet illustrated six boxes all of the same size are used. These are screwed together through their sides and ends which come into

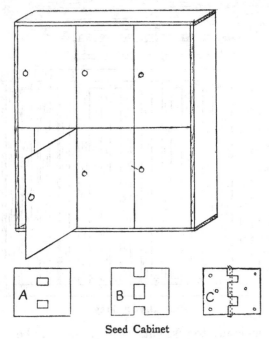

Seed Cabinet

contact. The lids have simple hinges made by the scholars. If necessary some of the boxes may have shelves fitted. The cabinet may stand either on its side or end. In the sketch a surrounding framework of wood, about $\frac{3}{8}$ inch thick, is shown—part of an old packing case would answer for this. If required a door may be made to fit the front of the cabinet, which can then be arranged to lock up. A similar type of cabinet could be made with the drawers of stout cardboard or of light woodwork.

A simple hinge may be made from an old tin. Two pieces are cut as at *A* and *B*. Each piece is then doubled, the parts fitted together and a piece of wire put through. Holes are then bored and the hinge is complete as at *C*.

A SEED CUPBOARD

A box may be made or bought and shelves fitted as suggested or altered to suit individual requirements.

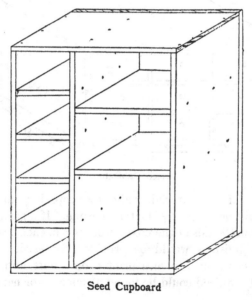

Seed Cupboard

10—2

This cupboard might have a door, which could be fitted with lock and key. The door could be made of strips with cross pieces nailed or screwed on, or the strips could be held together by four strips made in the shape of a square standing on an edge. This exercise could also be carried out in light woodwork. The cupboard illustrated measures $24'' \times 18'' \times 8''$.

SEED CABINET

A rather more elaborate form of seed cabinet is shown than the one previously described. The drawers may be made of cigar boxes. The lids should be taken off first and the end pieces planed level with the sides. The partitions and shelves carrying the drawers may

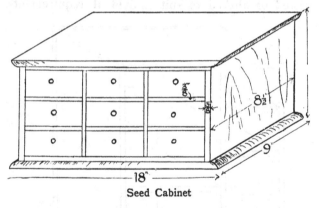

Seed Cabinet

be made of $\frac{3}{8}''$ material, while the top, bottom and sides could be made of $\frac{5}{8}''$ matchboarding. If a partition, made from the lids of the boxes, were put in each drawer, then 18 divisions would be provided for seeds. The drawers could have names printed on fronts, and thus any seed required could be found at once. For handles,

screw rings similar to those used in the back of picture
frames would answer. Careful work in cutting and
fitting the interior framework is required to make a nice
cabinet.

A Pea Guard or Seed Protector

A pea guard is an object that may be varied considerably in form. A square, triangular, or oblong piece
of wood will answer equally as well as a circular or
elliptical one. Sometimes two sticks are crossed so
that with the ground line they form a triangular shape
and cotton is fastened to them. A piece of hazel or
willow, bent to shape, will also answer all practical

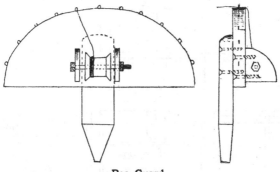

Pea Guard

purposes. But while pea guards may be more or less
roughly made, a well-finished pair, made in the manual
class, forms an excellent model. After trueing up the
wood, omitting gauging and planing to width, an ellipse
$10\frac{1}{2}'' \times 12''$ or other suitable size should be marked out.
After the ellipse has been cut out and finished with the
spokeshave it can be sawn, lengthwise of the wood, into
halves. Staples or nails can be used for fastening the

cotton to. They should be placed about 1½ inches apart and worked from each end towards the middle. The part that holds the protector in the ground calls for no comment. In the example shown two brackets are fixed, between which a reel of cotton will go so as to simplify threading. A wing nut and bolt forms a good arrangement for holding the cotton reel.

A Tool Shed

A tool shed is best built in sections. The measurements given may of course be altered as necessary. The

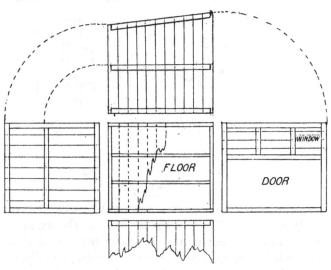

FLOOR

WINDOW

DOOR

Development of Tool Shed

base or floor is 6′ × 5′ and the joists are 3″ × 2″. The sides are bolted to the end pieces and the four pieces are then bolted to the floor section. The floor is of $\frac{7}{8}''$, and the sides and roof of $\frac{5}{8}''$, matchboarding. The roof can be covered with asphalt or other suitable material. The door is of matchboard and fitted with lock and key or staple and padlock. When completed it is coated with a preservative.

BRACKETS

Two brackets are shown, one made of mild steel and the other of wood. The metal one should be designed

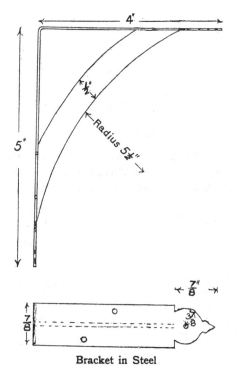

Bracket in Steel

by the pupil. Brackets are useful in the tool shed for the fixing up of shelves.

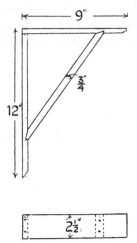

Bracket in Wood

Tool Rack

Racks of some kind for storing the tools should be made. These may be of different forms. One is illustrated made from a length of wood about three inches

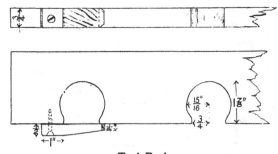

Tool Rack

wide by three-quarters of an inch thick. Holes, as shown, to suit sizes of handles of tools, are cut with a frame saw. The tools are held in place by means of buttons. To fasten the rack in position a strip of wood is first nailed to the shed and the rack is then nailed to this.

HAT PEGS

A few hat or coat pegs should be put in the tool shed and these can be made by the scholars, either of wood or metal. They are too well known to need illustrating.

A PROPAGATOR

A propagator is a most useful contrivance for raising tender plants, etc., requiring a temperature somewhere about 60° Fah. It consists of two parts, a base and a frame.

The base is a rectangular box without top or bottom, measuring about $2' \times 1\frac{1}{2}' \times 1\frac{1}{2}'$. An old packing case might be utilised for this purpose. It should be made with the pieces close fitting, or else there will be a loss of heat. The top of the box is covered with perforated zinc, the edges of which are turned over the sides of the box. A doorway is made to which a sliding door is fitted. Around the top of the box are nailed strips of wood, as shown, which will project about $1\frac{1}{2}$ inches above the top and thus form a part to hold the top frame in position. Before fixing on the frame suspend an old lamp shade or a piece of sheet iron (a discarded frying pan with handle taken off will answer admirably) about two inches below the perforated zinc, so as to come just over the top of the lamp when in position. By this device the heat will be better diffused.

The upper part is a frame with a hinged lid. This hinged lid is glazed. The frame, when not in use as

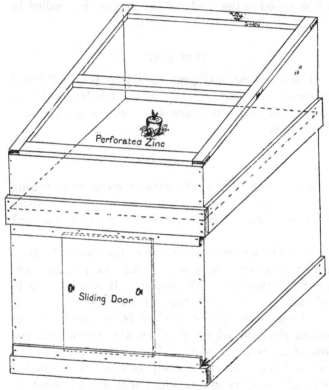

Perforated Zinc

Sliding Door

Plant Propagator

part of the propagator, may be used in the garden as an ordinary frame.

To use. The propagator should be placed in a shed or cool greenhouse. A small lamp, with a large reservoir for oil, is placed inside the box compartment. The door can be opened so as to regulate the temperature.   On the perforated zinc is placed about 4 inches of coconut fibre, and into this are plunged the pans of seeds, seedlings, etc.   It is advisable to keep in the frame a small vessel with water for watering purposes, as well as to maintain a damp atmosphere.

Light Rustic Woodwork

Very artistic-looking small objects can be made of natural sticks covered with "silver paper."   Stands for

Light Rustic Woodwork

cut flowers, etc. for a table or master's desk are easily made. Rough sketches are given indicating the type of object possible. For bowl, a half coco-nut shell answers admirably, while glass vessels that have contained potted meats, etc. also do nicely. These too may be covered with "silver paper." If a piece of wire netting can be cut to fit the top of the vessel in each case, then flowers can be held in much better than would otherwise be the case. The "silver paper" should be glued to the sticks. Work of this kind would make a useful break to the ordinary light woodwork lessons. Judgment and appreciation of artistic effect must be brought into play in the making and finishing of these objects.

Pillar for Rambler Roses

The propagation of rambler roses from cuttings is quite simple, and no school garden should be without a specimen or two of such roses as Crimson Rambler and Dorothy Perkins.

A rustic pillar can be made of three uprights; larch with the branches not closely trimmed to the posts is a suitable wood, but silver birch, etc. poles may also be used. The poles are sunk in the ground and then nailed together as shown. Seven feet above the ground would be a convenient height.

If wood is used from which the bark has been removed, then the structure should be given a coating of some preservative such as carbolineum or creasote.

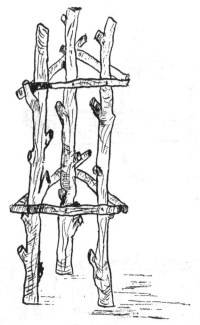

Rose Pillar

A Rustic Garden Seat

A rustic seat for the playground or school garden can readily be made of some branches of apple or other rustic-looking wood. The seat may be made of strips of yellow deal planed up and then coated with a preservative. The shape of the seat would be governed somewhat by the wood available. The rough sketches given may be taken as suggestive only. To nail the arms, etc. strong wire nails 4 to 6 inches long are necessary, while at times parts may be bound together by means of pieces of hoop iron.

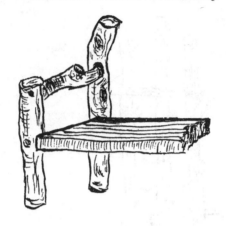

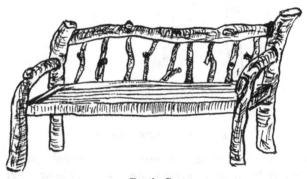

Rustic Seat

RUSTIC WINDOW BOX

A rustic window box is an object that lends itself to very many different ways of treatment. A box may be made with tiles in the front, but a rustic-looking one is much more effective, and can as a rule be more easily made and at a smaller cost. In any case a box of required dimensions should first be made, the bottom of which should have a number of holes to allow of drainage. The box may be covered with strips of larch, silver birch, virgin cork, etc. Endeavour so to clothe

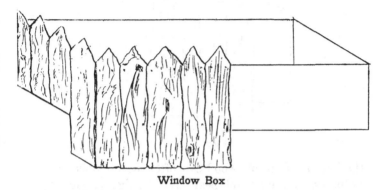

Window Box

the box as to make it look as natural as possible. Note that the end strips are cut shorter so as to allow the box to fit on the window ledge. The box may be of yellow deal, one inch thick.

Rustic Flower Box and Stand

A rustic flower box and stands are illustrated which would be suitable for standing in the school or in the garden. The box should first be made of stout wood and the bottom piece pierced by a few holes for drainage. The sides may then be covered with larch, silver birch, virgin cork, or other suitable material. The sides should be carried a little lower down than the bottom of the box so as to hide the top part of the stand.

The stand is a simple affair made of a few stout branches of almost any rustic-looking wood. Two to

three inches thick is quite stout enough. The more
branching, moss-covered and rustic-looking the wood

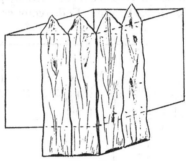

Plant Box

the better. Suitable wood can often be obtained from
prunings of apple and pear trees, hedge maple, etc.
The legs are nailed to each side of the top, and cross pieces

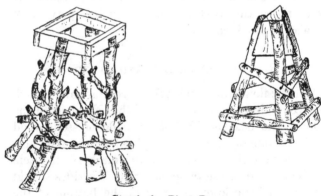

Stands for Plant Box

are also nailed to the legs. These pieces should be as
crooked as possible and put on quite irregularly. Hand

and eye rather than the rule must be largely employed in making this object. Instead of a square top for the stand a triangular one can be made and then three legs only will be required, and these can often be arranged better than four.

Plant Box for School

This plant box would be suitable to stand on a table in school or on a window ledge. The sides may be nailed or screwed together or dovetailed. If teak wood is used the box may be nicely made and finished off and so left,

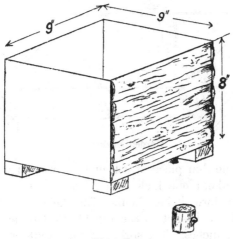

Plant Box

but if yellow deal is employed then it would look better if covered with some rustic wood. Four feet are put on the bottom. These may be square blocks or simply very small lengths of a pole. Many other forms of plant boxes may suggest themselves to the reader.

GARDEN BASKET

A garden basket can be made in the following manner. The ends are made of yellow deal about half an inch thick, 10 inches long by 5½ inches deep. The body is made of a sheet of thin yellow deal, about an eighth of an inch thick, and is nailed to the ends through pieces of binding which may either be a narrow strip of the thin yellow deal or else a piece of thin sheet zinc. An extra strip may be fixed to the top edge of each side of the basket to help strengthen it. The handle is made

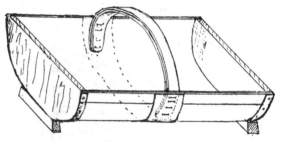

Garden Basket

either of one thin piece of ash or else two strips of the thin deal about one inch wide. The wood should be steamed before being bent. The height from the bottom of the basket to the top of the handle should be about 9 inches. The spring of the handle will keep the sides of the basket stiff, but if thought fit a wider piece of the thin yellow deal can be put round the outside of the basket. The handle is secured at each side by means of a few stitches of either thin wire or waxed string. To keep the basket off the ground two strips of wood can be nailed on to form feet.

CONCRETE EDGING FOR PATHS

An edging to the garden paths is often desired, but
the cost of tiles is likely to be prohibitive. Wood strips
if tarred may be used.   In many districts the making of
concrete edging could be undertaken without much
difficulty.   A pattern box or mould must first be made,
its internal dimensions and shape being those of the
required concrete block.   20 inches long, 8 inches deep

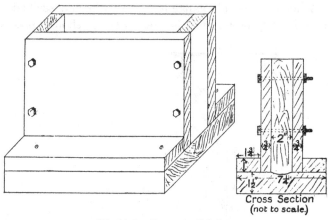

Cross Section
(not to scale.)

Mould for Concrete Edging

and 1½ to 2 inches thick would be a suitable size.   The
concrete would be made as follows :—Mix together on a
board 3 parts by measure of clean sharp sand and
1½ parts by measure of fine gravel.   Take 1 part Port-
land cement and sprinkle evenly over the sand and then
mix thoroughly by means of a shovel.   When this has
been done pour water over the heap from a water can
fitted with a rose.   Stir well and water until the whole
is of the consistency of thick mortar.   Fill a bucket

with the concrete and pour it into the moulds and leave to set. When this has taken place take the mould apart and the block should then be stacked with the others and all covered with wet bags, which are kept wet until the blocks have thoroughly hardened. Plant pots, boxes and other similar articles can likewise be made of concrete.

A Rain Gauge

A home-made rain gauge has already been described and a drawing of a funnel is now given. It should, pre-

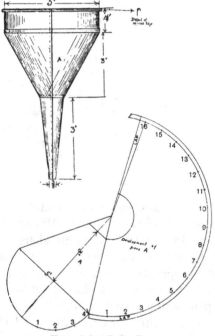

Funnel for Rain Gauge

ferably, be made of copper. Full details are given in
the illustrations. A cylinder can also be made of copper
of a size to take the funnel, or a glass bottle may be
used as a collecting vessel.

Hoe

Among metalwork objects capable of being made
by scholars, a hoe may be named. Drawings of one are

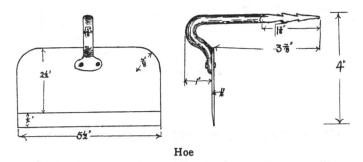

Hoe

given and modifications may be made where deemed
advisable. It is made of mild steel.

A Garden Trowel

An elevation of a garden trowel is shown as well as
a development of the blade and a section across it. The
blade is of mild steel about $\frac{1}{16}$ of an inch thick. The
tang is of mild steel $\frac{3}{8}$-inch diameter. A wooden handle
could be made and fitted. This might be hexagonal in
section and taper from the top towards the ferrule end.

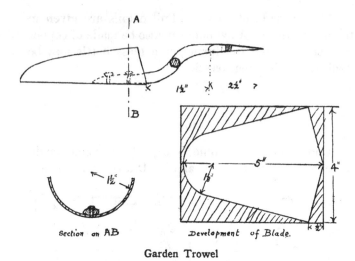

Garden Trowel

Garden Reels in Wrought Iron or Mild Steel

Where metalwork is possible a pair of garden reels can be made of iron or mild steel. Two lengths of iron rod of $\frac{3}{8}$-inch diameter are used to make the pins. These have the tops bent into ring shapes, while the other ends are pointed, being left either square or round in section. One of the pins should have a reel part which will revolve around the pin to which it is affixed. This part may be made of sheet iron about $\frac{5}{8}$-inch wide. Two uprights are riveted through these bars. Holes are next drilled through the bars so that the pin will go through them. To keep the reel in position a small hole is drilled through the pin just below the carrier, and through this a piece of wire may be driven.

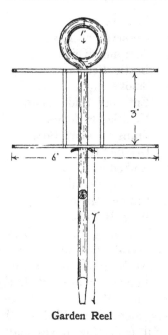

Garden Reel

BEE-KEEPING APPARATUS AND APPLIANCES

Apparatus and appliances connected with bee-keeping can be made by the more advanced scholars in the manual class. Some of the objects, such as a beehive, would be undertaken as collective work. It will be noted that the objects are constructed of wood, metal, and other materials. They should only be made as models when they are afterwards intended to be used *by the scholars* at school or at home. In such cases a beekeeper's book with illustrations or the actual objects will be available to work from. It has in consequence not been thought fit to give drawings. Among the objects that can be made are the following :—

1. Beehive.
2. Observatory Hive.
3. Scraping Knife.
4. Bottle Feeder.
5. Smoker.
6. Entrance Slide Door.
7. Ventilator Cones.
8. Transit Honey Box.
9. Driving Irons.
10. Embedder.
11. Wiring Board.
12. Frame Carrier.
13. Wax Moulds.
14. Travelling Box for Bees.

Drawings of several other objects might have been given, but they have been omitted chiefly because they are already well known or are not likely to present any difficulty in the making. A shelf to be used for potting purposes might be added to the outside of tool shed. Useful collecting boxes may be made of empty cigar boxes if the bottoms are fitted with cork lino. A full-sized garden frame or a wheelbarrow could be undertaken as collective work. A steelyard or balance for weighing (though not for sale) garden produce might be made, while a foot rest for a garden seat and stands for flower-pots may be made. A fumigating apparatus for a greenhouse and an edge cutter or edging iron for trimming the edges of grass plots may be cited as additional exercises in metalwork.

EXERCISES

The following questions are to be considered as suggestive. It is not intended that the scholars should work through the whole or any part of them.

1. Make sketches of a parsnip seed and write a description of it.

2. Describe how you would grow parsnips. Treat of the preparation of the soil, the planting of the seed, and the after-cultivation of the plants. What would you do with the crop when fully grown ?

3. Describe the cultivation of onions (*a*) for ordinary purposes, (*b*) for pickling. Name variety or varieties you would grow.

4. When would you sow seeds to produce early spring cabbage ? Describe the treatment you would give the young plants to produce fine sturdy specimens. What could you do in spring to help the cabbage to heart quickly ?

5. Why cannot beet, parsnips and carrots be transplanted with success as a general rule ?

6. How would you get seed of early potatoes from a potato patch ? How would you treat the seed until planting time the following year ? Give your reasons.

7. Name any vegetables that are hardy enough to withstand the winter.

8. When can you plant broad beans ? Describe how you plant them and their after treatment. What pest does the broad bean suffer from ? Describe it and say what treatment you would adopt to prevent its ravages.

9. Which method of culture would you prefer to adopt in raising onions (*a*) autumn sown, (*b*) sown under glass, (*c*) sown out of doors where they have to grow ? Give reasons for your answer.

10. Which vegetables are earthed up and why ?

11. Where, in the garden, would you grow lettuce and how would you obtain a succession throughout the summer ?

12. Which plants are blanched before they are fit for use and how is this done in the case of each plant you mention ?

13. Name three herbs grown in the school garden and state how each is propagated. What is the particular use of each?

14. Describe the cultivation of leeks. Make a drawing of a leek.

15. How would you protect newly sown peas from birds?

16. What bird or birds attack growing peas? What do you do to prevent their ravages?

17. Which plants especially do you find slugs and snails assail? What steps do you take to prevent their ravages?

18. For what purpose do thrushes visit the garden and what traces do they often leave behind them of their visit?

19. Name any bird that you frequently see in the garden and give as full a description of it as possible. Do you consider it a friend or a foe?

20. Why do you earth up potatoes? Draw a diagrammatic section, crossways of a row, through a row of potatoes.

21. Why is the ground made firm for an onion bed?

22. State fully how you raise, grow, and treat celery.

23. Do you know any vegetable that can be used instead of turnips and having somewhat the same flavour? How are turnips grown? What do turnips suffer from in their early stages of growth and what steps would you take to prevent such attacks?

24. What state should the ground be in for planting small seeds and why?

25. How do you sow carrot seed? Try to describe a carrot seed.

26. Into what two classes may carrots be divided? Say what you can of the differences between the two classes. How would you obtain a succession of young carrots throughout the summer and autumn?

27. Describe the cultivation of vegetable marrows.

28. State what you consider the chief points of excellence in a pair of vegetable marrows intended for exhibition.

29. How do you stick peas? Which is the more profitable variety to grow—tall or dwarf growing peas?

30. What steps would you take to obtain peas for seed from your own plants?

31. From which vegetables can you readily obtain seed for the following year? Name any precautions you would take to ensure seed of good quality.

32. Make a drawing of three radishes tied together in a bunch. Describe the cultivation of radishes.

33. If you have a garden frame, tell some of the purposes for which it is used.

34. Name half a dozen weeds that can usually be found in the school garden. Describe one of them fully and draw it.

35. In what ways do weeds propagate themselves? Why are weeds so difficult to eradicate?

36. Describe fully the culture of either dwarf or runner beans.

37. Make a drawing of a bunch of kidney beans.

38. Make a drawing of a kidney bean pod opened and showing the seed.

39. Why are runner beans more prolific through being constantly picked? Do you know any other plant or plants that give better results when treated in a similar manner?

40. What pests do onions suffer from while growing and what would you do to prevent or check the damage? Have you been successful in your efforts in this direction?

41. Draw and describe an onion.

42. Cut a "chible" onion through the centre longways and make a drawing and write an account of what you see.

43. What facts do you try to keep in mind when planning your garden plot and why?

44. What do you mean by rotation of crops? Name crops that should be moved to fresh ground each year and some that will do on the same ground for longer periods. Why is this?

45. What do you do with garden refuse? Why?

46. Describe the making of a garden bonfire. Of what use is it to the garden?

47. At what time of the year do you put manure on your garden? State your reasons for this practice.

48. Which vegetable crops should not be planted on freshly manured ground and why ?

49. Make a drawing of a beetroot. Describe the culture of beet.

50. For which vegetable crops in particular should you dig the ground deeply ? Explain why.

51. In judging a dish of onions for *quality* what points would be looked for ?

52. Name different varieties of "winter greens" and state when you usually sow the seed and the after treatment of the crops.

53. What crop or crops may follow early potatoes ?

54. What do you mean by "puddling" in plants ?

55. Explain any steps you would take to prevent, as far as possible, the garden being infected with potato disease.

56. Why are lime, soot, wood ashes used on the garden ?

57. How could you obtain a supply of beet or onion seed from your own plants ?

58. Make drawings and describe the two kinds of blossoms that grow on a vegetable marrow plant.

59. What advantages are there in digging up the ground as early in the winter as possible ?

60. Name the points looked for in judging a dish of potatoes at a flower show.

61. It is said "Shallots should be planted on the shortest day and raised on the longest day." What does this statement mean ?

62. Make a drawing of a bunch of ripened shallots.

63. Describe the culture of brussels sprouts.

64. Is it right to aim at producing as many crops from the land as possible ? State any steps you have taken to carry this idea into effect.

65. Name any crops that are grown for use as salads. Describe the culture of one of these.

66. Describe, with drawings, either a dandelion or a groundsel plant.

67. Write a description of either couch grass or shepherd's purse. Make a drawing of the one you select.

68. Name any climbing plants grown in the garden. How is the habit provided for in the case of each of the plants you name?

69. Why are "twiggy" sticks used for peas and straight sticks for kidney beans?

70. Say all you can about savoys and their culture.

71. How do you get rid of surface weeds? Of deep-rooted weeds?

72. How do you keep the garden paths free of weeds? Could this be done in any other way?

73. Name any tool that you would use to help keep clean a weedy patch of ground. Make a memory sketch of the tool and describe it.

74. Which green crops will stand winter weather most successfully? Describe fully the cultivation of one of these crops.

75. Why are spring broccoli planted with their heads facing north and in firm ground?

76. Make a drawing of a bunch of peas and write a description of a pea pod.

77. What trouble do you have with cabbage during the summer months? Describe the pest and state what steps you take to stop its ravages.

78. Syringe thoroughly a summer cabbage bed infested with grub with a strong solution of salt and water, and write a description of the results noticed upon (a) the grubs, (b) the cabbage.

79. Describe any experiment in plant growth you are doing in the school garden.

80. You intend to show a collection of six vegetables at a flower show. State as fully as you can what vegetables you would select and what would be the qualities that a judge should look for in each.

81. If you have a small garden plot, which is the more profitable to grow and why—Potatoes or other vegetables?

82. How would you prepare the ground preparatory to planting winter greens, brussels sprouts, etc.? Why should the ground be made firm for these plants? Would Dutch hoeing of this firm ground after the crops have become established be a disadvantage or otherwise?

83. Draw and describe a sprouted seed pot to.

84. Write a history of any vegetable plant you know.

85. Write a description and give sketches of a spade.

86. Explain, with sketches, the mechanical principles involved in the use of a spade or fork for digging.

87. When would you use a spade and when a fork in digging? State reasons for the choice.

88. Make freehand sketches of a Dutch hoe and of a draw hoe.

89. Explain the mechanical principle involved in the use of a wheelbarrow or a handbarrow.

90. Which garden tools in use act as levers and which as wedges?

91. What kind of wood is usually employed in making the shafts of spades, forks, hoes, etc. and why? Describe this wood and say all you can about it, and the reasons for its selection for these purposes.

92. Write a description of an ash tree and make sketches illustrating the tree in a summer and a winter state respectively.

93. Make a series of drawings of an ash twig to show development of buds.

94. Make a drawing of an ash leaf and also of ash "keys."

95. For what purposes is the garden rake employed?

96. Describe fully the construction of a garden rake. Is it better to have the handle rather long or rather short and why?

97. Which garden tool is the most useful to employ in hot dry summer weather and why? Describe it.

98. Make drawings to scale of a garden reel.

99. For what purposes is a dibber used? State any disadvantages arising from its use in particular cases.

100. If told to clean a garden path of weeds, which tool or tools would you select for the purpose and why?

101. Make sketches of and describe a garden trowel and a small handfork.

102. Make drawings showing front and side views of a spade and a Dutch hoe. Project the one view from the other.

103. Give a full description of a garden fork, especially stating of what materials it is made.

104. For what purposes is a potato hoe required ? Name any other tool you have seen used for a similar purpose.

105. Write a list of the school garden tools and give a full description of one of them.

106. Explain how you use the spade in digging.

107. Write a description of a Dutch hoe and state how it is used.

108. Which tool would you use to edge your garden plot and how would you use it ?

109. If you have a watering can or a garden syringe, make drawings of them.

110. Describe the construction of a garden syringe.

111. Can you say anything of the early forms of any of the garden tools ? Name any countries or peoples still using primitive tools for cultivating the land.

112. After using garden tools what do you do to them before storing away and why ?

113. Describe your tool shed.

114. Make scale drawings of one of the garden tools using the metric system of measurement. (Each scholar to have, as far as possible, a separate tool to draw.)

115. Make a drawing and give a description of a potato box for storing seed potatoes.

116. Of what use is a garden basket ? Describe one and note especially those features which fit it for this use.

117. Make a drawing of a garden basket.

118. Write a list of the tools that you consider absolutely necessary for use in a garden. State if you could make any of these tools yourself and which.

119. Which would you select, a spade with a blade about 10 inches deep and 7 inches wide, or one with a blade about 12 inches deep and 6 inches wide ? Give reasons for your choice.

120. What tool or tools do you use to prune gooseberry and currant bushes ? Describe and make drawings of same.

121. Suppose you wish to purchase a spade ; what points would you look for in making a selection ? About what price would you pay ?

122. Write to a firm of tool merchants (at a distance from the place where you live) a request for their tool catalogue.

123. Write an imaginary letter ordering a tool, enclosing remittance, and stating how and by what route the tool is to be sent.

124. Which garden tool do you best like to use and why ?

125. Make a drawing to scale of the tool shed.

126. Draw to a convenient scale the garden wheelbarrow or handbarrow.

127. Name any garden tool you could make and state how to make it and from what materials.

128. I have some posts carrying railings in my garden and when digging around these during the winter I found on the south side of each a number of cabbage moth chrysalids. There were none on the north side. While still in the grub stage did they seek the warmer side of the posts, or was there any other reason ? If you have an opportunity try to investigate this matter.

129. Make drawings illustrating how you would bud a rose brier. Describe the process.

130. Can you layer a carnation ? If so, make sketches showing how you would do it.

131. Name any flowers that you propagate by means of cuttings, and describe how you make a cutting and get it to root.

132. Name any perennial flowers that you grow in the school garden and write a description of one of these.

133. Name some bulbous flowers that you are fond of and state what you can about how they are planted, when they flower, and their after treatment.

134. Describe the culture of Sweet Williams or Canterbury Bells.

135. Name any " annual " flowering plants you would like to grow.

136. State what flowers you would choose for a position on a rockery.

137. What position is suitable for growing ferns ? Why ?

138. Name a few flowers that you would like to grow in a border so as to give bloom through as long a period of the year as possible.

139. Which flowers do the bees seem to visit most frequently ? Give any reasons why you think they should have any preference.

140. Have you seen flowers visited by any other insects than hive bees ? If so describe the blossom and its visitor.

141. What pests have you noticed on a rose bush and how would you treat these to get rid of them ?

142. Have you ever noticed a rose bush with many of its leaves that have had little round pieces cut out of their edges each about the size of a threepenny piece ? Try to find out what does the mischief and also what is done with the pieces.

143. If you have any mullein plants growing in the garden keep a look out on them for grubs, and then note the colouring of (*a*) grub, (*b*) the plant in blossom.

144. If you find a butterfly or a moth at rest on a fence, wall, etc. look at the colour of the creature and compare this with the colour of its surroundings. If there is any similarity, try to account for it. Keep watch and try to find out other examples of the same thing.

145. If you have plants infested with aphides, try to find out what other creature or creatures visit them and what takes place. Try to find out the life-story of any such visitors. Illustrate by drawings.

146. Would you describe ants as being useful or otherwise in a garden and why ?

147. Give drawings from nature to illustrate the life-story of a cabbage white butterfly, an onion fly, a raspberry shoot moth, a chafer beetle.

148. Describe fully one or more centipedes found in the garden.

149. How many different kinds of snails can you find in the school garden ? Make drawings of each.

150. Can you make any suggestions for improving the flower border of the school garden ?

151. Describe the culture of sweet peas.

152. How would you plant flowers that are annuals ? When the plants are well up what would you do ?

153. Make drawings and write descriptions of either a Shirley poppy or an Iceland poppy.

154. Write a description of a crocus plant.

155. When in the garden see if you can find an insect known as a "hover fly." What other insect does it closely represent and for what reason ? Watch it and try to find out what you can of its doings. Investigate its life-story.

156. Examine every caterpillar of a cabbage butterfly you find, and if you notice any that look rather ill, keep them for a time and see if later they die and in their bodies appear a number of grubs or chrysalids. If so, keep these latter and see what they turn into. Find out the life-story of these creatures.

157. Expose to frost and cold weather for a time some chrysalids and see how many of these mature, and compare with other chrysalids you have kept that were not so exposed to the weather.

158. Find out the effect of frost and weather upon a large lump of soil, and hence the effect of digging up the ground rough in autumn or early winter.

159. Test on several occasions the temperature at a depth of one inch of (a) soil that has been well Dutch hoed, and (b) soil that has not been so stirred.

160. Make an analysis of the soil of your garden to show the proportion of clay and grit.

161. Describe how you would trench the soil and how you would double dig it. Draw diagrams to illustrate each process.

162. Make drawings to scale of how you would place broad beans in a trench.

163. Find an earwig and after killing it, try to open its wings. Make a drawing of a wing. Would the creature be better named "ear-wing" and if so, why ? (You should use a magnifying glass in making the drawing.)

164. What kind of food do earwigs take and when do they feed? How does this suggest a means of capturing them ? What plant in particular are they fond of attacking ?

165. Find a woodlouse and make a drawing of it and also write a description of it. Where did you look for it and why?

166. One of the prettiest and most handsome of the flies that can be seen in the garden is the gauzy green lacewing fly. Try to find a specimen of this creature and interest yourself in finding out all you can of its earlier stages of existence.

167. Watch wasps in the garden and see what they visit. If they visit a spider's web, try to see why and therefore try to find out if wasps are in any way serviceable.

168. You may be able to find a frog or a toad in the school garden. If so find out where it hides, and as much as possible concerning what it is doing in the garden.

169. Why are earthworms considered to be friends of the gardener?

170. Name any creatures that feed upon the earthworm. How do they capture their prey?

171. A dead mole or similar creature might be left lying on the ground in the summer time and watched to see if it is visited by some beetles. If so, watch what they do with the body and find out why.

172. Watch a garden spider catching and disposing of its prey and afterwards write a description of what you have noticed.

173. How could you propagate raspberries, and when established how would you treat them year after year?

174. Describe the cultivation of strawberries. Why have they this name?

175. Name any shrubs that would be suitable for growing on the school premises. State whether they are evergreen or otherwise.

176. How could you raise a stock of euonymus bushes, of laurel, of cotoneaster, of golden privet?

177. How could you raise a stock of laurestinus bushes?

178. What plant is sometimes used as a garden edging? How is it treated?

179. Make to scale a cardboard model of the tool shed and its fittings.

180. Measure, and find the area in metric measurement of the catchment vessel of the rain-gauge. Measure and find likewise the area of the inside of the measuring glass and hence find out the relationship between the two.

181. Describe an experiment you might conduct to find out approximately the amount of water given off by a cabbage in the course of 24 hours, and hence calculate the amount of water that would be given off from an acre of cabbages planted 18″ apart each way.

182. How would you treat kidney or runner beans which you were unable to stick?

183. Weigh the onions obtained from your plot; find out the area of ground on which they grew, and hence calculate the weight of a similar crop from an acre of land.

184. Which vegetable crop that you grow do you consider the most profitable? Give reasons for your answer.

185. Give the name of any vegetable grown in the school garden which you have never tasted when cooked.

186. What attention, if any, would you give to rhubarb?

187. Name the herbs which are grown in the school garden and state for what purposes each is used.

188. What is the direction of the slope of your garden and what influence has this, if any, on the production of crops?

189. How would you raise pansy or violas from (a) seed, (b) cuttings? How would you treat the plants when they began to get "straggly"?

190. Describe how you would raise forget-me-nots and polyanthus from seed.

191. Make drawings from nature some time during the winter months of shoots of black currants and twigs of nuts respectively (a) showing shoot or twig with normal buds, (b) ditto with big bud. From a suitable copy make an enlarged drawing of a big bud mite.

192. Say how you prepare a gooseberry cutting for planting.

193. Describe how you make a black currant cutting.

194. How do you graft an apple tree? Describe fully and state why you graft the tree.

195. Name a weed which, in quantity, can be dug in to act as "green manure." Describe this weed.

196. Collect a few of the very finest blackberries you can get and then select the very finest one for seed. Sow the seed in the school garden, and cultivate the plant with the view to showing whether improvement takes place in its fruit production over ordinary wild ones.

197. Make drawings of gooseberry, red currant and black currant cuttings prepared for planting.

198. What ideas would you keep before you when pruning a gooseberry bush (a) one year old, (b) five years old? Why?

199. Describe how to prune a black currant bush.

200. Why are grease bands put round the stems of fruit trees in the winter? Describe as fully as you can.

201. Have you seen fruit trees sprayed? If so, what was done and with what purpose in view?

202. Why do you like gardening?

From the scale drawings of the school garden work out the following sums:

203. Find area of the garden. Express your answer in (1) sq. yards, (2) sq. perches, (3) acres.

204. What area do the plots allotted to the scholars occupy?

205. How much of the school garden is devoted to paths? Express your answer as (a) fraction, (b) decimal, (c) percentage of the whole. What is the area of the paths?

206. Find the number of feet in length of all the garden paths.

207. Find the cost of asphalting the main garden path at a cost of 1½d. per square foot.

208. Find the cost of wooden bordering to all the paths at 3d. per yard run of material.

209. What is the perimeter of the garden?

210. What is the area of the largest circular-shaped piece of ground that could be made out of the school garden?

211. Find the cost of erecting and tarring a fence round the garden at a cost of 3s. a yard run, tarring, both sides, being 3d. a yard run extra.

212. Find the cost of building a brick or stone wall round two sides of the garden from details given you by your teacher.

213. How many rows of potatoes could you plant on your plot allowing the rows 3 feet apart? What would be the total length of all these rows?

214. If it takes 4 wheelbarrowfuls of manure for your plot, how many loads would this be per acre, assuming that a load of manure is equivalent to 15 barrow loads?

215. Find the area of the paths round your plot.

216. Find the cost of trenching the garden at 1*s.* 8*d.* per square pole.

217. If 1 cubic foot of rain water weighs 1000 oz., find the weight of water that fell on your plot, when your rain-gauge registers ·25 inch rainfall.

218. If land in your neighbourhood is worth £5 an acre per annum, what would be the rent for one year at this rate (*a*) of the school garden, (*b*) of your plot?

219. What is the length and breadth of your plot in metric measurement? Use these figures and find the area.

220. Onions are worth 1*s.* a peck of 14 lbs., and a peck is produced from 3 rows across your garden plot. Find the value of the onion crop that would cover your plot. What return would an acre of land yield at the same rate?

221. If 15 per cent. of the onions (details as in 220) are destroyed by maggot, what loss would this represent in a crop, one and a half acres in extent?

222. If a cubic yard of your garden soil weighs one ton, find the approximate weight of the soil of your plot above the subsoil.

223. Write a general description of the honey bee.

224. What do you know about wax, pollen, and propolis?

225. Make drawings of a worker bee and drone.

226. Name some of the advantages of using moveable combs.

227. Describe, as fully as possible, a modern beehive and its parts.

228. Say all you can about queen bees.

229. Why do bees swarm ? What steps, if any, would you take to prevent swarming ? How would you take a swarm of bees, and what would you do with them ?

230. Write all you can about the cell-making instinct of bees.

231. In what position would you place a hive in the garden and why ?

232. " A swarm of bees in May is worth a load of hay." Explain fully this saying.

233. What enemies and pests attack bees ? Make drawings of as many of these as possible, and describe any steps you would take to prevent their ravages.

234. What large moth is sometimes found visiting beehives ? For what reason ?

235. Write a description of a smoker and describe how and why it is used.

236. Make a drawing of a smoker.

237. Say all you can about subduing and handling bees.

238. What are the disadvantages of a skep hive ?

239. How would you start an apiary ?

240. When and why do bees need feeding ? What food is given to bees ?

241. Name any varieties of bees you know and describe, as far as you are able, the characteristics of each.

242. Write an account of foul brood and state any steps you would take to prevent or get rid of it.

243. What do you know about the "Isle of Wight disease"? Say all you can about wintering bees.

244. How is honey extracted ?

245. What is comb foundation ? How is it made ? Why is it used ?

246. When a bee visits a flower describe fully what it is likely to carry away. How did it get these things, by what means are they carried, and what use will be made of them ?

247. Write accounts as fully as you can of the uses of weeds, and also of harm done by weeds.

248. Name any weeds characteristic of damp and wet soils and situations.

249. Name weeds that are characteristic of good soils.

250. What part of garden work do you like best ? Give your reasons.

HELPS

Books likely to be useful

Lessons on Soil, by Dr Russell; Cambridge University Press. 1/6 This book deals with simple experimental work that can be carried out in most schools. It is one of the Cambridge Nature Study books and there are several other volumes in the series that might be useful to the teacher.

Hand and Eye Training. Its principles and methods, by H. Holman; Pitman & Sons. 3/-. This book should be read by every teacher.

Our Country's Birds, by W. J. Gordon; Simpkin, Marshall & Co. 6/-. There are several more books in this series dealing with Mammals, Fish, Butterflies, Moths, etc. The books are serviceable for identification purposes.

The Romance of Wild Flowers, by Edward Step; F. Warne & Co. 6/-. The title sufficiently explains the object of the book.

Text Book of Elementary Botany, by C. Laurie; Alman & Son. 2/6. A work of a practical and experimental character.

Shown to the Children books, T. C. & E. C. Jack, contain some works that would be welcome in any school library. *The Look About You Nature Study Books*, by the same firm, are good and helpful.

A Nature Study Guide, by Furneaux; Longmans & Co. 3/6. A valuable book dealing with the practical side of nature study.

Lessons on Plant and Animal Life, by Dr Rennie ; University Tutorial Press. 4/6. Deals with plant and animal life, and contains just the kind of information which scholars should be taught to observe. It is a teacher's book for school use.

Helps

Text Book of Agricultural Entomology, by E. A. Ormerod; Simpkin, Marshall & Co. 3/6. A useful reference work dealing with small animal life.

Exercises in Nature Study. Stages I, II, III. Jas. Nisbet. 6*d.* each. By means of questions a large number of interesting and useful nature study problems are given.

Land and Freshwater Shells, of British Isles, by Ed. Rimmer; John Grant & Co., and *Land and Freshwater Shells,* Swan Sonnenschein & Co. Two books dealing with slugs and snails.

Injurious and Useful Insects, by L. C. Miall; G. Bell & Sons. 3/6. A useful work which deals more particularly with the anatomy of insect life.

Wild Bees, Ants, and Wasps, by E. Saunders; Routledge.

First Books of Science. Macmillan & Co. Zoology, 1/-. Botany, 1/-. Chemistry of the Garden, Cousins, 1/-. Horticulture, Dean, 1/-. Geology for Beginners, Watts, 2/6. Useful books for teachers.

Principles of Horticulture, by W. M. Webb; Blackie & Co. 2/-. Worth a place in any teacher's library.

The Story of the Plants, by Grant Allen; Newnes. 1/-. A charming little book, dealing with non-technical botany from an evolutionary standpoint. It deals with heredity, variation, natural selection and adaptation to environment.

Moths. Vols. I and II. South. Warne & Co. 7/6 each.
Butterflies.   „   „   6/-.
Splendidly illustrated works and up to date.

Weeds. Long. Smith, Elder & Co. 6/-.

Principles of Educational Woodwork, by Binns and Marsden. G. Bell & Sons. 5/-.

Plant Study in School, Field and Garden. Bridges and Dicks. Ralph, Holland & Co. 3/6.

Board of Agriculture Leaflets. Bound volumes of these are recommended.

Memorandum on Nature Study and the Teaching of Science in Scottish Schools. (Cd. 4024.) 3*d.* Full of useful suggestions.

Suggestions on Rural Education, by T. S. Dymond, 1905. 3*d.*

Memorandum on Principles and Methods of Rural Education. 1911. 3*d.*

Suggestions for the Teaching of Gardening. 1910. (Circular 746.) 1*d*.

Home Bottled Fruits and How to do Them. Harvey & Healing, Cheltenham. 1/-. A practical manual on fruit bottling.

Rules for Judging at Flower Shows, Royal Horticultural Society. 1/6.

The Marketing of Garden Produce, by R. L. Castle; John Lane. 2/6.

Origin and History of our Garden Vegetables. Rev. Prof. Henslow; Royal Hort. Society. 2/-.

Beetle Collector's Handbook.

Among the many helps a teacher should have at his command books must necessarily take a leading place. A reference library, either small or great, according as circumstances will allow, is a necessity. The above list of books has been given with the idea of indicating some that are likely to prove useful. A complete or exhaustive list has not been attempted. Public and other libraries, where available, should be made use of where possible. In many of our towns Public Libraries with Reference Departments exist, and where such is the case, it is a splendid practice to send scholars to such a department to find out any information required.

Books, however, while of great help cannot take the place of personal observation and work carried out by an enthusiastic teacher. The best knowledge must always be that which is achieved through *doing*, whether by teacher or by pupil, and this can only be gained by working through the subject one's own self, either alone or with the scholars.

Teachers keen on educational progress would be well rewarded by attending a Summer School such as is held at Scarborough, Ambleside, etc. It does not much matter which subject is studied, whether woodwork, cardboard modelling, brushwork, or nature study. It is the inspiring and suggestive hints and ideas that are likely to prove most helpful, for these are likely to bear upon any or all of the subjects of the school curriculum. A month spent at a Summer School inspires and renews enthusiasm.

In conclusion the teacher may well take to heart and cherish the motto:

Disce ut doceas, "Learn that ye may teach."

INDEX

Printed in the United States
By Bookmasters